Engineering Materials 3

by the same author

Engineering Materials 2

Engineering Materials 3

W. Bolton

Heinemann Professional Publishing

Heinemann Professional Publishing Ltd
Halley Court, Jordan Hill, Oxford OX2 8EJ

OXFORD   LONDON   MELBOURNE   AUCKLAND

First published 1988

© W. Bolton 1988

British Library Cataloguing in Publication Data
Bolton, W. (William), *1933–*
  Engineering materials 3.
  1. Materials
  I. Title
  620.1′1

ISBN 0 434 90139 3

Typeset in Great Britain by
Scarborough Typesetting Services
and printed by
Robert Hartnoll Ltd, Bodmin

Contents

Preface vii

1 Mechanical properties of metals 1
Mechanical properties 1
The tensile test 1
Hardness measurements 9
Impact tests 14
Toughness 18
Fatigue tests 19
Long-term behaviour – creep 22
Which mechanical property? 24
Problems 26

2 Mechanical properties of polymeric materials 29
Polymeric materials 29
Mechanical properties 29
The tensile test 30
Creep behaviour of plastics 33
Hardness measurements 34
Impact tests 34
Fatigue properties of plastics 35
Service requirements 35
Problems 36

3 Alloying of metals 39
Alloys 39
Iron alloys 39
Copper alloys 41
Mixtures, solutions and compounds 42
Solubility and precipitation 42
Phase 43
Alloy types 44
Equilibrium diagram 44
Problems 50

4 Ferrous alloys 52
The iron–carbon system 52
Alloy steels 56
Critical change points 58
Heat treatment of steel 59
Hardenability 64
Heat treatment equipment 68
Problems 69

5 Non-ferrous alloys 71
Aluminium 71
Aluminium alloys 72
Copper 76
Copper alloys 77
Magnesium 82
Magnesium alloys 82
Nickel 83
Nickel alloys 83
Titanium 83
Titanium alloys 84
Zinc 84
Zinc alloys 84
Comparison of non-ferrous alloys 85
Precipitation hardening 87
Problems 88

6 Polymeric materials 91
Polymers 91
Structure of polymers 92
Glass transition temperature 94
Effects of temperature and time on mechanical properties 95
Temperature and polymer use 96
Orientation 96
Additives 98
Copolymers 100
Common thermoplastics and their properties 101
Thermosetting polymers 106
Elastomers 108
Polymer foams 111
Polymer based sandwich materials 111
Polymer composites 111
Problems 112

Index 114

Preface

This book has been written to provide, for the engineering student:

- an overall view of the test methods used with metals and polymeric materials, and of the interpretation and validity of their results
- a basic understanding of the alloying of metals
- a basic understanding of carbon steels and their heat treatment
- a basic understanding of polymeric materials, their structures and properties and the factors affecting those properties
- an awareness of the materials commonly used in engineering, their properties and applications.

The aim has been to make this as far as possible a self-contained textbook, so that the reader does not need to refer to the earlier book in this series, *Engineering Materials* 2. Thus some of the content of that book is repeated here. The book more than covers the BTEC unit Engineering Materials III (U84/266), to some extent because of the deliberate inclusion of aspects of the earlier unit but also because certain topics from other BTEC units at level III are covered in order to give a more comprehensive overview of engineering materials.

The book uses much of the material that appeared in the sequence of books *Materials Technology* 2, 3 and 4, designed to cover the earlier Technican Education Council units, but this has been completely reorganised, and in parts rewritten, to cover the new BTEC units.

1 Mechanical properties of metals

Objectives: At the end of this chapter you should be able to:

Interpret the results of tensile testing, hardness measurements, impact testing, fatigue testing and creep measurements.
Select the relevant property data for a given service requirement for a component.

MECHANICAL PROPERTIES

The following are some of the main terms used in describing mechanical properties.

Elasticity. A material is said to be elastic if it returns to its original shape and dimensions when straining forces are removed.

Plasticity. A material is said to show plasticity if it retains a permanent deformation after straining forces are removed. Metals are generally elastic at low stresses but can show plasticity at higher stresses. High temperatures can reduce the stress at which plasticity occurs, hence the wide use of hot working for metals.

Strength. This is the ability of materials to withstand gradually applied forces without rupture. The forces may be tensile, compressive or shear. An important point in any consideration of strength is whether the situation is one of gradually applied forces, impact or shock loading, or alternating stresses where the stress is repeatedly reversed.

Toughness. This is the ability of a material to withstand shock loads.

Fatigue strength. This is the ability of a material to withstand alternating stresses where the stress is repeatedly reversed.

Ductility. A ductile material will show a considerable amount of plastic deformation before rupture. This allows such materials to be worked and formed into the required shapes by manipulation.

Brittleness. A brittle material will break without noticeable plastic deformation. Brittleness is the opposite of ductility.

Malleability. This is the capacity of a material to undergo deformation by hammering or squeezing, e.g. rolling. It is a similar property to ductility.

Hardness. This can be defined as the ability of a material to withstand surface deformation, indentation or abrasion.

THE TENSILE TEST

In a tensile test, measurements are made of the force required to extend a standard size test piece at a constant rate, the elongation of a specified gauge length of the test piece being measured by some form of extensometer. A force–extension graph can be produced from the data.

Figure 1.1 shows a typical force–extension graph for a ductile material such as mild steel. The following are the various stages occurring during the tensile test giving such a graph.

2 Mechanical properties of metals

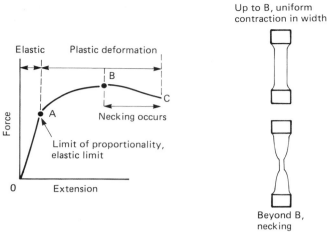

Figure 1.1 Force–extension graph for a ductile metal

1 For the part of the force–extension graph OA the extension is directly proportional to the force. *Hooke's law* is said to be obeyed.

2 Point A is called the *limit of proportionality*.

3 Up to a certain force, the material will return to its original shape and dimensions when the force is removed. The material is said to be elastic. The maximum force for which a material is elastic is called the *elastic limit*. For many materials the elastic limit and the limit of proportionality are almost identical.

4 Higher forces than the limit of proportionality result in a considerable increase in extension, the extension no longer being proportional to the force. The material stretches more easily.

5 Beyond the elastic limit the material will not return to its original dimensions when the force is removed, the material retaining a permanent change in dimensions. This is referred to as a *permanent set* or *plastic deformation*.

6 As the force is gradually increased beyond the elastic limit the extensions increase more rapidly until a point is reached (B) when without any further increase in force there is an increase in extension. When this occurs the test piece begins to show a pronounced 'neck' or region of local contraction. Further extensions can be produced with a decreasing force until a point (C) is reached when the material breaks.

Stress and strain Tensile or compressive *stress* is defined as the force per unit area acting on the material to cause a change in length. It has the unit of N/m^2 or Pa (pascal) with $1\,Pa = 1\,N/m^2$. *Strain* is defined as the extension per unit length and has no unit since it is a ratio.

$$\text{stress} = \frac{\text{force}}{\text{area}} \qquad \text{strain} = \frac{\text{extension}}{\text{original length}}$$

During a tensile test the cross-sectional area of the material changes, particularly when necking occurs. However, to avoid

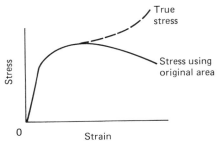

Figure 1.2 True stress

having to measure the cross-sectional dimensions of the test piece continually during a tensile test, the stress is normally taken as

$$\text{stress} = \frac{\text{force}}{\text{original area}}$$

This stress will differ from the true stress because of changes in the cross-sectional area; however, because most materials are usually used well below the elastic limit, when the area change is very small, the true stress will differ very little from the stress calculated on the basis of the original area.

When necking occurs, the extension increases although the force is decreased, as in *Figure 1.1*. The stress calculated on the basis of the original area must also decrease, since it is proportional to the force. However, the true stress increases because there is a decrease in area over which the force is applied. A graph of true stress against strain thus differs from a graph of stress against strain when the stress is calculated on the basis of the original area (*Figure 1.2*).

Data from tensile tests

Figure 1.3 shows the type of stress–strain graph that occurs with a low carbon steel. The following are data taken from such a graph.

1 The slope of the stress–strain graph up to the limit of proportionality (*Figure 1.4*) is called the *tensile modulus* or *Young's modulus*. The modulus has the unit N/m² or Pa.

$$\text{Tensile modulus} = \frac{\text{stress}}{\text{strain}}$$

2 For a tensile stress–strain graph the term *tensile strength* is used for the maximum stress the material withstands (*Figure 1.3*).

3 Beyond the limit of proportionality the strain is no longer proportional to the stress and in some materials a situation may arise where the strain continues to increase without any increase in stress, i.e. the material continues to extend without any increase in force. The material is said to be yielding and the stress at which it occurs is called the *yield point*. For the graph shown in *Figure 1.3* there are two yield points, called the upper and lower yield stresses. Where data are given as yield points, with no distinction between upper and lower points, the values given are for the lower yield points. After the yield points have been passed an increase in stress is necessary for an increase in strain.

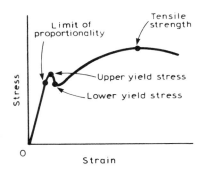

Figure 1.3 Stress–strain graph for a low carbon steel

4 Many materials do not have well-defined yield points, their stress–strain graphs being of the form shown in *Figure 1.5*. In such instances a *proof stress* is specified rather than a yield stress. The 0.2% proof stress is defined as that stress which results in a 0.2% offset, i.e. the stress given by a line drawn parallel to the linear part of the graph and passing through the 0.2% strain value. Similarly the 0.1% proof stress is given by drawing a line on the stress–strain graph parallel to the linear part of the graph and passing through the 0.1% strain value.

5 After the test piece has broken, the pieces are fitted together and the final gauge length is measured. With the initial gauge length, this enables a quantity called the *percentage elongation* to be calculated.

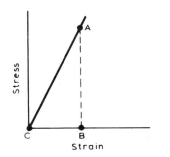

Figure 1.4 Young's modulus (tensile modulus) equals AB/BC

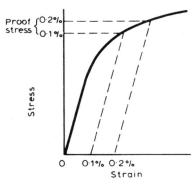

Figure 1.5 Determination of proof stress

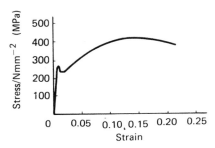

Figure 1.6 Stress–strain graph for a sample of mild steel

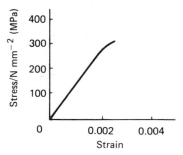

Figure 1.7 Stress–strain graph for a sample of cast iron

$$\text{percentage elongation} = \frac{\text{(final length)} - \text{(initial length)}}{\text{initial length}} \times 100$$

Similarly, another quantity called the *percentage reduction in area* can be determined from the initial cross-sectional area and the smallest cross-sectional area at fracture:

$$\text{percentage reduction in area} = \frac{\text{(final area)} - \text{(initial area)}}{\text{initial area}} \times 100$$

Example
Figure 1.6 shows the stress–strain graph for a sample of mild steel. Determine (a) the limit of proportionality, (b) the upper yield point, (c) the lower yield point and (d) the tensile strength.
   (a) The limit of proportionality is about 240 N mm^{-2} or MPa.
   (b) The upper yield point is about 280 N mm^{-2} or MPa.
   (c) The lower yield point is about 240 N mm^{-2} or MPa.
   (d) The tensile strength is about 400 N mm^{-2} or MPa.

Example
Figure 1.7 shows the stress–strain graph for a sample of cast iron. Determine Young's modulus for the sample.
   Young's modulus is about 125 kN mm^{-2} or GPa.

Example
Figure 1.8 shows the stress–strain graph for a sample of an aluminium alloy. Determine (a) the 0.1% proof stress, (b) the 0.2% proof stress.
   (a) The 0.1% proof stress is about 460 N mm^{-2} or MPa.
   (b) The 0.2% proof stress is about 520 N mm^{-2} or MPa.

Example
A sample of mild steel had an initial gauge length of 69 mm. After breaking, the gauge length had become 92 mm. What is the percentage elongation?

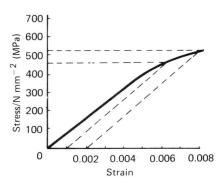

Figure 1.8 Part of the stress–strain graph for a sample of an aluminium alloy

$$\text{Percentage elongation} = \frac{\text{(final length)} - \text{(initial length)}}{\text{initial length}} \times 100$$

$$= \frac{92 - 69}{69}$$

$$= 33\%$$

The tensile test piece

In order to eliminate any variations in tensile test data due to differences in the shapes of test pieces, standard shapes are adopted. *Figure 1.9* shows the forms of two standard test pieces, one being a flat test piece and the other a round test piece. The following are the dimensions of some standard test pieces:

Flat test pieces

b/mm	L_0/mm	L_c/mm	L_t/mm	r/mm
25	100	125	300	25
12.5	50	63	200	25
6	24	30	100	12
3	12	15	50	6

Round test pieces

A/mm^2	d/mm	L_0/mm	L_c/mm	r/mm Wrought material	r/mm Cast material
200	15.96	80	88	15	30
150	13.82	69	76	13	26
100	11.28	56	62	10	20
50	7.98	40	44	8	16
25	5.64	28	31	5	10
12.5	3.99	20	21	4	8

(The information in the above tables was extracted from BS 18).

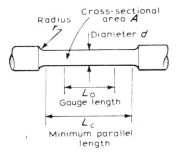

(a) Round test piece

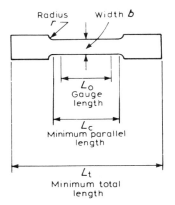

(b) Flat test piece

Figure 1.9 Tensile test pieces

For the tensile test data for the same material to give essentially the same stress–strain graph, regardless of the length of the test piece used, it is vital that the above standard dimensions be adhered to. An important feature of the dimensions is the radius given for the shoulders of the test pieces. Variations in the radii can affect markedly the tensile test data. Very small radii can cause localised stress concentrations which may result in the test piece failing prematurely. The surface finish of the test piece is also important for the same reason.

The round test pieces are said to be *proportional test pieces*, for which the relationship between the gauge length L_0 and the cross-sectional area of the piece A is specified in the relevant British Standard as being

$$L_0 = 5.65\sqrt{A}$$

With circular cross-sections $A = \frac{1}{4}\pi d^2$ the relationship becomes, to a reasonable approximation,

$$L_0 = 5d$$

The reason for the specification of a relationship between the gauge length and the cross-sectional area of the test piece is in order to give reproducible test results for the same test material when

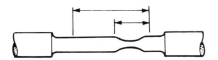

Figure 1.10 Percentage elongations for the two different gauge lengths differ considerably

different size test specimens are used. When a ductile material is being extended in the plastic region of the stress–strain relationship, the cross-sectional area of the piece does not reduce uniformly but necking occurs. The effect of this is to cause most of the further plastic deformation to occur in the necked region where the cross-sectional area is least. The percentage elongation can thus differ markedly for different gauge lengths encompassing this necked portion of the test piece (*Figure 1.10*). Doubling the gauge length does not double the elongation because most of the elongation is in such a small part of the gauge length. The same percentage elongation is however given if

$$\frac{\text{gauge length}}{\sqrt{(\text{cross-sectional area})}} = \text{a constant}$$

In England the constant is chosen to have the fixed value of 5.65.

Validity of tensile test data

The tensile test piece is usually chosen so that its properties are indicative of the properties of a component or components. How valid is the data obtained from a test piece?

1 The properties of a component may not be the same in all parts of it. Due to different cooling rates in different parts of a casting (e.g. the surface compared with the core, or thin sections compared with thick sections) the internal structure of the material may differ and hence different tensile properties exist in different parts. A tensile test piece cut from one part may thus not represent the properties of the entire casting. For the same reason, the properties of a specially cast test piece may not be the same as those of a casting because the different sizes of the two result in different cooling rates.

2 The properties of a component may not be the same in all directions. Wrought products, i.e. products produced by manipulative methods, tend to show directionality of properties. Thus, for example, with rolling the tensile properties in the longitudinal, transverse and through thickness directions will differ. The following table illustrates this for rolled brass strip (70% copper, 30% zinc).

Angle to rolling direction	Tensile strength /N mm^{-2} or MPa	Percentage elongation
0°	740	3
45°	770	3
90°	850	2

3 The temperature in service of the component may not be the same as that of the test piece when the tensile test data was obtained. The tensile properties of metals depend on temperature. In general the tensile modulus and the tensile strength both decrease with an increase in temperature. The percentage elongation tends to increase with an increase in temperature (*Figure 1.11*).

4 The rate of loading of a component may differ from that used for the test piece. The data obtained from a tensile test are affected by the rate at which the test piece is strained, so in order to give standardised results, tensile tests are usually carried out at a con-

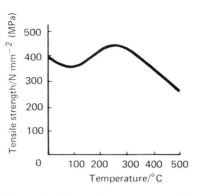
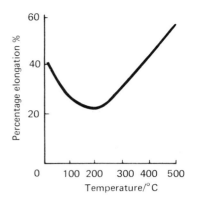

Figure 1.11 Data for a carbon steel

stant strain rate. For metals tested at room temperature, the standard rate specified is BS 18 is a strain of 0.5 per minute or less.

Typical tensile test results   The following table gives typical tensile strength and tensile modulus values for metals.

Material	Tensile strength /N mm^{-2} (MPa)	Tensile modulus /kN mm^{-2} (GPa)
Aluminium alloys	100 to 550	70
Copper alloys	200 to 1300	110
Magnesium alloys	150 to 350	45
Nickel alloys	400 to 1600	200
Titanium alloys	400 to 1600	100
Zinc alloys	200 to 350	100
Grey cast iron	150 to 400	100
Mild steel	350 to 500	200
Ferritic stainless steel	500 to 600	200
Martensitic stainless steel	450 to 1300	200

Ruling section   If you look up the mechanical properties of a steel in the data supplied by the manufacturer or other standard tables you will find that different values of the mechanical properties are quoted for different limiting ruling sections. The *limiting ruling section* is the maximum diameter of round bar at the centre of which the specified properties may be obtained. Here is an example of such information:

Steel	Condition	Limiting ruling section/mm	Tensile strength /N mm^{-2} (MPa)	Minimum elongation %
070 M 55	Hardened and tempered	19 63 100	850 to 1000 777 to 930 700 to 850	12 14 14

The reason for the difference of mechanical properties for different size bars of the same steel is that during the heat treatment different rates of cooling occur at the centres of such bars due to their differences in sizes. This results in differences in microstructure and hence differences in mechanical properties.

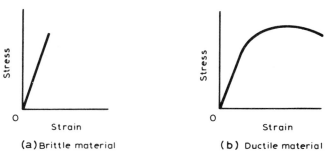

Figure 1.12 Stress–strain graphs to fracture

Interpreting tensile test data

Figure 1.12 shows the types of stress–strain graphs produced by brittle and ductile materials. *Brittle* materials show little plastic deformation before fracture, *ductile* materials show a considerable amount of plastic deformation. If you drop a china tea cup and it breaks you can stick the pieces together again and still have the same tea cup shape. The material used for the tea cup is a brittle material and little, if any, plastic deformation took place prior to fracture. If the wing of a motor car is involved in a collision it is likely to show considerable deformation rather than a fracture. The mild steel used for the car body work is a ductile material.

A brittle material will show only a small percentage elongation, i.e. the length of the broken specimen is very little different from the initial length as little plastic deformation has occurred. The material also shows little, if any, sign of necking. A ductile material, however, shows quite a large percentage elongation and quite significant necking.

Grey cast iron is a brittle material, it has a percentage elongation of about 0.5 to 0.7%. Mild steel is a reasonably ductile material and has a percentage elongation of the order of 30%.

The tensile modulus of a material can be taken as a measure of the stiffness of the material. The higher the value of the modulus the stiffer the material, i.e. the greater the force needed to produce a given strain within the limit of proportionality region (*Figure 1.13*). Mild steel has a tensile modulus of about 200 kN mm^{-2} (GPa) while an aluminium alloy may have a modulus of 70 kN mm^{-2} (GPa). A strip of mild steel is thus stiffer than a corresponding strip of aluminium alloy.

The strength of a material under low rates of loading is indicated by its tensile strength. An alloy steel may have a strength as high as 1500 N mm^{-2} (MPa), while an aluminium alloy may only have a strength of 200 N mm^{-2} (MPa). The steel is obviously much stronger than the aluminium alloy.

Figure 1.13 Stress–strain graphs within the limit of proportionality region

Example
Which of the materials shown in *Figure 1.14* is (a) the most ductile, (b) the most brittle, (c) the strongest, (d) the stiffest?
 (a) Material C, (b) Material B, (c) Material A, (d) Material A.

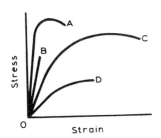

Figure 1.14

Example
Which of the following materials is the most ductile?

Material	Percentage elongation %
80–20 brass	50
70–30 brass	70
60–40 brass	40

The most ductile material is the one with the largest percentage elongation, i.e. the 70–30 brass.

Example
Figure 1.11 shows how the tensile strength and percentage elongation varies with temperature for a carbon steel. At what temperature is the steel (a) strongest, (b) least ductile?

(a) The steel is strongest at a temperature of about 250°C, within the range for which the data is given.

(b) The steel is least ductile at a temperature of about 200°C, within the range for which the data is given.

HARDNESS MEASUREMENTS

The *hardness* of a material may be specified in terms of some standard test involving indentation or scratching of the surface of the material. Hardness is essentially the resistance of the surface of a material to deformation. There is no absolute scale for hardness, each hardness form of test having its own scale. Though some relationships exist between results on one scale and those on another, care has to be taken in making comparisons because the tests associated with the scales are measuring different things.

The most common form of hardness tests for metals involves standard indentors being pressed into the surface of the material concerned. Measurements associated with the indentation are then taken as a measure of the hardness of the surface. The Brinell test, the Vickers test and the Rockwell test are the main forms of such tests.

With the *Brinell test* (*Figure 1.15*) a hardened steel ball is pressed for a time of 10 to 15 s into the surface of the material by a standard force. After the load and the ball have been removed the diameter of the indentation is measured. The Brinell hardness number (signified by HB) is obtained by dividing the size of the force applied by the spherical surface area of the indentation. This area can be obtained,

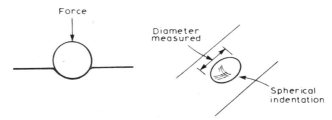

Figure 1.15 The basis of the Brinell hardness test

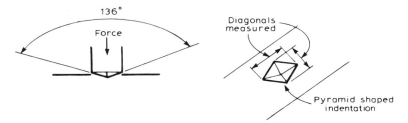

Figure 1.16 The basis of the Vickers hardness test

either by calculation or the use of tables, from the values of the diameter of the ball used and the diameter of the indentation.

$$\text{Brinell Hardness number} = \frac{\text{applied force}}{\text{spherical surface area of indentation}}$$

The units used for the area are mm² and for the force are kgf (1 kgf = 9.8 N). The British Standard for this test is BS 240.

The diameter D of the ball used and the size of the applied force F are chosen, for the British Standard, to give F/D^2 values of 1, 5, 10 or 30, the diameters of the balls being 1, 2, 5 or 10 mm. In principle, the same value of F/D^2 will give the same hardness value, regardless of the diameter of the ball used.

The Brinell test cannot be used with very soft or very hard materials. In the one case the indentation becomes equal to the diameter of the ball and in the other case there is either no or little indentation on which measurements can be based. The thickness of the material being tested should be at least ten times the depth of the indentation if the results are not to be affected by the thickness of the sample.

The *Vickers test* (*Figure 1.16*) uses a diamond indenter which is pressed for 10 to 15 s into the surface of the material under test. The result is a square-shaped impression. After the load and indenter are removed the diagonals of the indentation are measured. The Vickers hardness number (HV) is obtained by dividing the size of the force applied by the surface area of the indentation. The surface area can be calculated, the indentation is assumed to be a right pyramid with

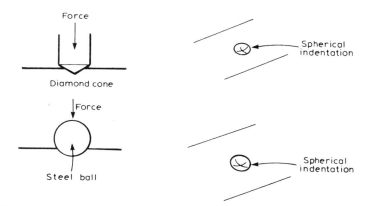

Figure 1.17 The basis of the Rockwell hardness test

a square base (the vertex angle of the pyramid is assumed to be the same as the vertex angle of the diamond, i.e. 136°), or obtained by using tables and the diagonal values. The relevant British Standard is BS 427.

The Vickers test has the advantage over the Brinell test of the increased accuracy that is possible in determining the diagonals of a square as opposed to the diameter of a circle. Otherwise it has the same limitations as the Brinell test.

The *Rockwell test* (*Figure 1.17*) uses either a diamond cone or a hardened steel ball as the indenter. A force is applied to press the indenter in contact with the surface. A further force is then applied and causes an increase in depth of the indenter penetration into the material. The additional force is then removed and there is some reduction in the depth of the indenter due to the deformation of the material not being entirely plastic. The difference in the final depth of the indenter and the initial depth, before the additional force was applied, is determined. This is the permanent increase in penetration (e) due to the additional force.

Rockwell Hardness number (HR) = $E - e$

where E is a constant determined by the form of the indenter. For the diamond cone indenter, E is 100, for the steel ball, E is 130.

There are a number of Rockwell scales, the scale being determined by the indenter and the additional force used. The following table indicates the scales and the types of materials for which each are typically used.

Scale	Indenter	Additional force/kN	Typical applications
A	Diamond	0.59	Thin steel and shallow case-hardened steel.
B	Ball 1.588 mm dia.	0.98	Copper alloys, aluminium alloys, soft steels.
C	Diamond	1.47	Steel, hard cast irons, deep case-hardened steel.
D	Diamond	0.98	Thin steel and medium case-hardened steel.
E	Ball 3.175 mm dia.	0.98	Cast iron, aluminium alloys, magnesium alloys, bearing metals.
F	Ball 1.588 mm dia.	0.59	Annealed copper alloys, thin soft sheet metals, brass.
G	Ball 1.588 mm dia.	1.47	Malleable irons, gun metals, bronzes, copper-nickel alloys.
H	Ball 3.175 mm dia.	0.59	Aluminium, lead, zinc.
K	Ball 3.175 mm dia.	1.47	Aluminium and magnesium alloys.
L	Ball 6.350 mm dia.	0.59	Plastics.
M	Ball 6.350 mm dia.	0.98	Plastics.
P	Ball 6.350 mm dia.	1.47	
R	Ball 12.70 mm dia.	0.59	Plastics.
S	Ball 12.70 mm dia.	0.98	
V	Ball 12.70 mm dia.	1.47	

The relevant British Standard for Rockwell tests is BS 891. In any reference to the results of a Rockwell test the scale letter must be quoted. The B and C scales are probably the most commonly used for metals.

For the most commonly used indenters with the Rockwell test the size of the indentation is rather small. Localised variations of structure, composition and roughness can thus affect the results. The Rockwell test is however more suitable for workshop or 'on site' use as it is less affected by surface conditions than the Brindell or Vickers tests which require flat and polished surfaces to permit accurate measurements. The test, as indicated above, cannot be used with thin sheet. A variation of the standard test, called the *Rockwell superficial hardness test*, can be used however with thin sheet. Smaller forces are used and the depth of the indentation is determined with a more sensitive device as much smaller indentations are used. The initial force is 29.4 N instead of 90.8 N. The following are the scales given by this test.

Scale	Indenter	Additional force/kN
15–N	Diamond	0.14
30–N	Diamond	0.29
45–N	Diamond	0.44
15–T	Ball 1.588 mm dia.	0.14
30–T	Ball 1.588 mm dia.	0.29
45–T	Ball 1.588 mm dia.	0.44

Note: numbers with the scale letter refer to the additional force values when expressed in kgf units (1 kgf = 9.8 N).

Comparison of the different hardness scales

The Brinell and Vickers tests both involve measurements of the surface areas of indentations; the form of the indenters however is different. The Rockwell tests involve measurements of the depths of penetration of the indenter. Thus the various tests are concerned with different forms of material deformation as an indication of hardness. There are no simple theoretical relationships between the various hardness scales though some approximate, experimentally derived, relationships have been obtained. Different relationships hold for different metals.

The following table shows the conversions that are used between the different tests for steels.

Brinell value	Vickers value	Rockwell values	
		B	C
112	114	66	
121	121	70	
131	137	74	
140	148	78	
153	162	82	
166	175	86	4
174	182	88	7
183	192	90	9
192	202	92	12
202	213	94	14
210	222	96	17
228	240	98	20
248	248	102	24
262	263	103	26
285	287	105	30
302	305	107	32
321	327	108	34
341	350	109	36

Brinell value	Vickers value	Rockwell values B	C
370	392		40
390	412		42
410	435		44
431	459		46
452	485		48
475	510		50
500	545		52

Up to a hardness value of 300 the Vickers and Brinell values are almost identical.

There is an approximate relationship between hardness values and tensile strengths. Thus for annealed steels the tensile strength in N mm^{-2} (MPa) is about 3.54 times the Brinell hardness value, and for quenched and tempered steels 3.24 times the Brinell hardness value. For brass the factor is about 5.6, and for aluminium alloys about 4.2.

Example
An aluminium alloy (4.0% Cu, 0.8% Mg, 0.5% Si, 0.7% Mn) has a hardness of 45 HB when annealed and 100 HB when solution treated and precipitation hardened. What might be the tensile strengths of the alloy in these two conditions?

If the tensile strength is assumed to be the hardness value multiplied by 4.2 then the tensile strength in the annealed condition is 189 N mm^{-2} (MPa) and in the heat treated condition is 420 N mm^{-2} (MPa). The measured values were 180 N mm^{-2} (MPa) and 430 N mm^{-2} (MPa).

The Moh scale of hardness

One form of hardness test is based on assessing the resistance of a material to being scratched. With the *Moh scale* ten materials are used to establish a scale. The materials are arranged so that each one will scratch the one preceding it in the scale but not the one that succeeds it. The following is the list of the materials in this scale.

1 Talc
2 Gypsum
3 Calcspar
4 Fluorspar
5 Apatite
6 Felspar
7 Quartz
8 Topax
9 Corundum
10 Diamond

Thus Felspar will scratch apatite but not quartz. Diamond will scratch all the materials, while talc will scratch none of them.

Ten stylii of the materials in the scale are used for the test. The hardness number of a material under test is one number less than that of the substance that just scratches it. Thus, for instance, glass can just be scratched by felspar but not by apatite. The glass has thus a hardness number of 5.

14 Mechanical properties of metals

Typical hardness values

The following table gives typical hardness values for metals at room temperature.

Material	Hardness value
Aluminium, commercial, annealed	21 HB
hard	40 HB
Aluminium – 1.25% Mn alloy, annealed	30 HB
hard	30 HB
Copper, oxygen-free HC, annealed	45 HB
hard	105 HB
Cartridge brass, 70% Cu, 30% Zn, annealed	65 HB
hard	185 HB
Magnesium alloy – 6% Al, 1% Zn, 0.3% Mn, forged	65 HB
Nickel alloy, Monel, annealed	110 HB
cold worked	240 HB
Titanium alloy – 5% Al, 2.5% tin, annealed	360 HB
Zinc casting alloy A, as cast	83 HB
Grey cast iron	210 HB
White cast iron	400 HB
Malleable cast iron, Blackheart	130 HB
Carbon steel – 0.2% carbon, normalised	150 HB
Stainless steel, austenitic (304), annealed	150 HB
cold worked	240 HB

Figure 1.18 shows the general range of hardness values for the different types of materials when related to Vickers, Brinell, Rockwell and Moh hardness scales.

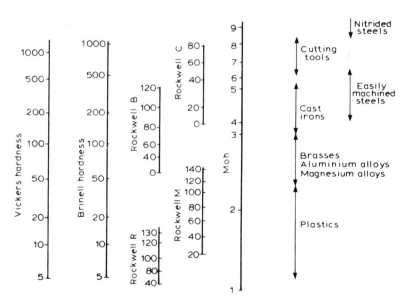

Figure 1.18 Hardness values

IMPACT TESTS

Impact tests are designed to simulate the response of a material to a high rate of loading, and involve a test piece being struck a sudden blow. There are two main forms of test, the *Izod* and *Charpy* tests. Both tests involve the same type of measurement but differ in the form of the test pieces. Both involve a pendulum swinging down

Mechanical properties of metals 15

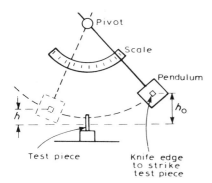

Figure 1.19 The principle of impact testing

from a specified height to hit the test piece (*Figure 1.19*). The height to which the pendulum rises after striking and breaking the test piece is a measure of the energy used in the breaking. If no energy were used the pendulum would swing up to the same height as it started from. The greater the energy used in the breaking the lower the height to which the pendulum rises.

With the Izod test the energy absorbed in breaking a cantilevered test piece (*Figure 1.20*) is measured. The test piece has a notch and the blow is struck on the same face of the piece as the notch is and at a fixed height above it. The test pieces used are, in the case of metals, either 10 mm square or 11.4 mm diameter if they conform to British Standards (BS 131: Part 1). *Figure 1.21* shows the details of a test piece conforming to British Standards. The British Standard test pieces for plastics (BS 2782: Part 3) are either 12.7 mm square or 12.7 mm by 6.4 to 12.7 mm depending on the thickness of the material concerned (*Figure 1.22*). With metals the pendulum strikes the test piece with a speed between 3 and 4 m/s, with plastics this speed is 2.44 m/s.

With the Charpy test the energy absorbed in breaking a beam test piece (*Figure 1.23*) is measured. The test piece is supported at each end and is notched in the middle between the two supports. The notch is on the face directly opposite to where the pendulum strikes the test piece. For metals the British Standard test piece has a square cross-section of side 10 mm and a length of 55 mm (BS 131: Parts 2 and 3). *Figure 1.24* shows the details of a standard test piece and the three forms of notch that are possible. The results obtained with the different forms of notch cannot be compared; for comparison purposes between metals, the same type of notch should be used. The test pieces for plastics are tested either in the notched or unnotched state. The notch is produced by milling a slot across one face, the slot of width 2 mm having a radius of less than 0.2 mm at the corners of the base and the walls of the slot. A standard test piece is 120 mm long, 15 mm wide and 10 mm thick in the case of moulded plastics. Different widths and thicknesses are used with sheet plastics. With metals the pendulum strikes the test piece with a speed between 3 and 5.5 m/s.

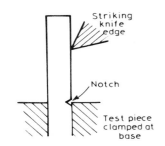

Figure 1.20 Form of the Izod test piece (elevation)

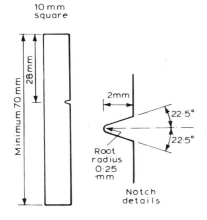

Figure 1.21 British Standard Izod test piece for a metal

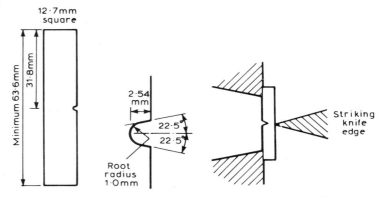

Figure 1.22 British Standard Izod test piece for a plastic

Figure 1.23 Form of the Charpy test piece (plan)

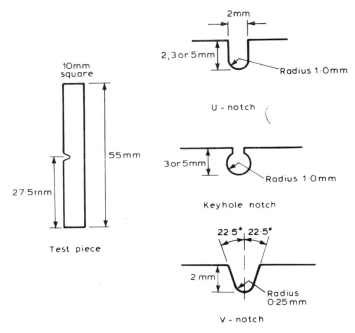

Figure 1.24 British Standard Charpy test piece for a metal

The results of impact tests need to specify not only the type of test, i.e. Izod or Charpy, but the form of the notch used. In the case of metals the results are expressed as the amount of energy absorbed by the test piece when it breaks. In some instances the results are expressed as the impact energy divided by the cross-sectional area of the specimen if not notched or by the area behind the notch for notched specimens.

Example

In an Izod test the pendulum in falling acquires a kinetic energy of 160 J before striking the test piece. If after fracturing the test piece the pendulum continues up to a height of 60% of that from which it started its motion, what is the energy used to fracture the piece?

The kinetic energy of the pendulum, after striking the test piece and breaking it, is transformed into potential energy. This potential energy is 60% of the potential energy with which the pendulum started. Thus 40% of the energy was used to break the test piece, i.e. 64 J.

Interpreting impact test results The fracture of materials can be classified roughly as either brittle or ductile fracture. With *brittle fracture* there is little, or no, plastic deformation prior to fracture. With *ductile fracture* the fracture is preceded by a considerable amount of plastic deformation. Less energy is absorbed with a brittle fracture than with a ductile fracture. Thus Izod and Charpy test results can give an indication of the brittleness of materials.

The appearance of the fractured surfaces after an impact test also gives information about the type of fracture that has occurred. With

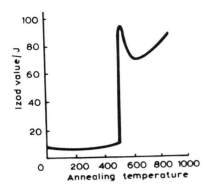

Figure 1.25 Effect of annealing temperature on Izod test values

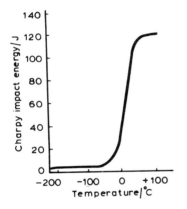

Figure 1.26 Effect of temperature on the Charpy V-notch impact energies for a 0.2% carbon steel

a brittle fracture of metals the surfaces are crystalline in appearance, with ductile fracture the surfaces are rough and fibrous in appearance. Also with a ductile failure there is a reduction in the cross-sectional area of the test piece, but with a brittle fracture there is virtually no change in the area. With plastics a brittle failure gives fracture surfaces which are smooth and glassy or somewhat splintered, with a ductile failure the surfaces often have a whitened appearance. With plastics the change in cross-sectional area can be considerable with a ductile failure.

One use of impact tests is to determine whether heat treatment has been successfully carried out. A comparatively small change in heat treatment can lead to quite noticeable changes in impact test results. The changes can be considerably more pronounced than changes in other mechanical properties, e.g. percentage elongation or tensile strength. *Figure 1.25* shows the effect on the Izod impact test results for cold worked mild steel annealed to different temperatures. The use of an impact test could then indicate whether annealing has been carried out to the correct temperature.

The properties of metals change with temperature. For example, a 0.2% carbon steel undergoes a gradual transition from a ductile to a brittle material at a temperature of about room temperature (*Figure 1.26*). At about −25°C the material is a brittle material with a Charpy V-notch impact energy of only about 4 J, whereas at about 100°C it is ductile with an impact energy of about 120 J. This type of change from a ductile to a brittle material can be charted by impact test results and the behaviour of the material at the various temperatures predicted.

Example
A structural steel gives a Charpy V-notch impact energy value of 27 J at room temperature. When the temperature drops to below 0°C this value decreases and at temperatures of −20°C it is considerably lower. Is the steel becoming more brittle or more ductile as the temperature falls?
The steel is becoming more brittle.

Typical impact test results

The following table gives typical impact test results for metals and plastics at 0°C.

Material	Charpy V Impact strength/J
Aluminium, commercial pure, annealed	30
Aluminium − 1.5% Mn alloy, annealed	80
hard	34
Copper, oxygen free HC, annealed	70
Cartridge brass (70% Cu, 30% Zn), annealed	88
¼ hard	21
Cupronickel (70% Cu, 30% Ni), annealed	157
Magnesium − 3% Al, 1% Zn alloy, annealed	8
Nickel alloy, Monel, annealed	290
Titanium − 5% Al, 2.5% Sn, annealed	24
Grey cast iron	3
Malleable cast iron, Blackheart, annealed	15
Austenitic stainless steel, annealed	217
Carbon steel, 0.2% carbon, as rolled	50

TOUGHNESS

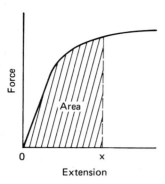

Figure 1.27 Work done in producing extension x = area under the curve up to extension x

The area under a force–extension graph for a material to a particular extension is the work done in stretching the material to that extension (*Figure 1.27*). The greater the area, the greater the amount of energy needed to extend the material. The maximum amount of energy a material can withstand without breaking is the area under the force–extension graph to the breaking point.

Work = force × extension

Since stress = $\dfrac{\text{force}}{\text{area}}$ and strain = $\dfrac{\text{extension}}{\text{length}}$

Work = (stress × area) × (strain × length)
 = (stress × strain) × (area × length)

Hence work per unit volume of material = stress × strain.

The area under a stress–strain graph is the energy per unit volume of material. Thus the maximum amount of energy per unit volume a material can withstand without breaking is the area under the stress–strain graph to the breaking point.

The term 'toughness' can be defined in a number of ways. It can be defined as the ability of a material to absorb energy during plastic deformation. With such a definition the area under the stress–strain graph can be taken as a measure of toughness. A consequence of this is that materials with high yield strength and high ductility, which hence must have large areas under their stress–strain graphs, must have high toughness.

An alternative definition, which is more widely used, is that toughness is the ability to withstand shock loads. The impact test involves shock loading and is thus taken as a measure of toughness. A tough material is thus one which requires a large amount of energy to break it when it is subject to a shock load.

Strength and toughness

The strength of a material is generally considered to be indicated by the tensile strength; the higher the tensile strength, the stronger the material. However the tensile strength is determined by a tensile test involving a low rate of loading. Impact tests give a better measure of 'strength' if a material is likely to be subjected to shock loads. The term 'toughness' can be used to describe the ability of a material to withstand shock loads. Thus the greater the energy absorbed in breaking a test piece in an impact test, the tougher the material.

Example
The following data show the mechanical properties of a number of carbon and carbon–manganese steels. Which steel has (a) the highest strength if static or a low rate of loading is involved and (b) the highest strength with shock loading, i.e. is the toughest material?

Mechanical properties of metals 19

Steel	Tensile strength /MPa	Yield stress /MPa	Elongation /%	Izod value /J	Brinell hardness number
080M40	540	280	16	20	150
150M19	540	325	18	40	150
120M28	540	325	16	34	150
150M28	590	355	16	34	170

All the above steels were in the normalised condition with a ruling section of 60.

(a) The highest tensile strength is shown by 150M28 and so this material has the highest strength for static or a low rate of loading.

(b) The highest impact value is given by material 150M19 and so this is the toughest material.

FATIGUE TESTS

Fatigue tests can be carried out in a number of ways, the way used being the one needed to simulate the type of stress changes that will occur to the material of a component when in service. There are thus bending-stress machines which bend a test piece of the material alternately one way and then the other (*Figure 1.28a*), and torsional-fatigue machines which twist the test piece alternately one way and then the other (*Figure 1.28b*). Another type of machine can be used to produce alternating tension and compression by direct stressing (*Figure 1.28c*).

The tests can be carried out with stresses which alternate about zero stress (*Figure 1.28d*), apply a repeated stress which varies from zero to some maximum stress (*Figure 1.28e*) or apply a stress which varies about some stress value and does not reach zero at all (*Figure 1.28f*).

In the case of the alternating stress (*Figure 1.28d*), the stress varies between $+S$ and $-S$. The tensile stress is denoted by a positive sign, the compressive stress by a negative sign; the stress range is thus $2S$. The mean stress is zero as the stress alternates equally about the zero stress. With the repeated stress (*Figure 1.28e*), the mean stress is half the stress range. With the fluctuating stress (*Figure 1.28f*) the mean stress is more than half the stress range.

In the fatigue tests, the machine is kept running, alternating the stress, until the specimen fails, the number of cycles of stressing up to failure being recorded by the machine. The test is repeated for the specimen subject to different stress ranges. Such tests enable graphs similar to those in *Figure 1.29* to be plotted. The vertical axis is the *stress amplitude*, half the stress range. For a stress amplitude greater than the value given by the graph line, failure occurs for the number of cycles concerned. These graphs are known as *S/N graphs*, the *S* denoting the stress amplitude and the *N* the number of cycles.

For the *S/N* graph in *Figure 1.29a* there is a stress amplitude for which the material will endure an indefinite number of stress cycles. The maximum value, S_D, being called the *fatigue limit*. For any stress amplitude greater than the fatigue limit, failure will occur if the material undergoes a sufficient number of stress cycles. With the *S/N* graph shown in *Figure 1.29b* there is no stress amplitude at

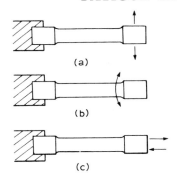
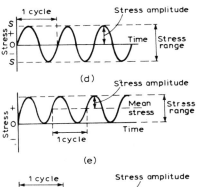
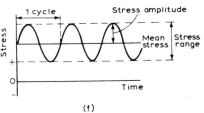

Figure 1.28 Fatigue testing (a) Bending, (b) Torsion, (c) Direct stress, (d) Alternating stress, (e) Repeated stress (f) Fluctuating stress

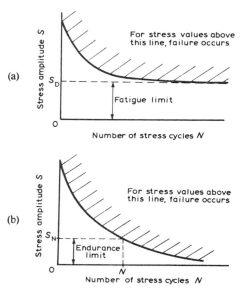

Figure 1.29 Typical S/N graphs for (a) a steel, (b) a non-ferrous alloy

which failure cannot occur; for such materials as *endurance limit* S_N is quoted. This is defined as the maximum stress amplitude which can be sustained for N cycles.

The fatigue limit, or the endurance limit at about 500 million cycles, for metals tends to lie between about a third and a half of the static tensile strength. This applies to most steels, aluminium alloys, brass, nickel and magnesium alloys. For example, a steel with a tensile strength of $420\,\text{Mn}\,\text{m}^{-2}$ (MPa) has a fatigue limit of $180\,\text{MN}\,\text{m}^{-2}$ (MPa), just under half the tensile strength. If used in a situation where it were subject to alternating stresses, such a steel would need to be limited to stress amplitudes below $180\,\text{MN}\,\text{m}^{-2}$ (MPa) if it were not to fail at some time. A magnesium alloy with a tensile strength of $290\,\text{MN}\,\text{m}^{-2}$ (MPa) has an endurance limit of $120\,\text{MN}\,\text{m}^{-2}$ (MPa), just under half the tensile strength. Such an alloy would need to be limited to stress amplitudes below $120\,\text{MN}\,\text{m}^{-2}$ (MPa) if it were to last to 500 million cycles.

Factors affecting the fatigue properties of metals

The main factors affecting the fatigue properties of a component are:

1. Stress concentrations caused by component design.
2. Corrosion.
3. Residual stresses.
4. Surface finish.
5. Temperature.

Fatigue of a component depends on the stress amplitude attained, the bigger the stress amplitude the fewer the stress cycles needed for failure. Stress concentrations caused by sudden changes in cross-section, keyways, holes or sharp corners can thus more easily lead to a fatigue failure. The presence of a countersunk hole was considered in one case to have lead to a stress concentration which could have

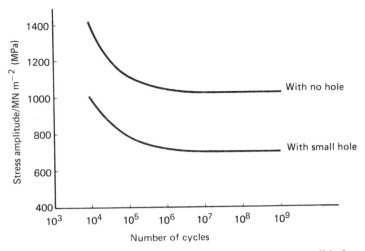

Figure 1.30 *S/N* graph for a steel both with and without a small hole acting as stress raiser. Note: number of cycles shown on logarithmic scale

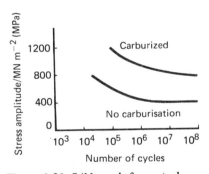

Figure 1.31 *S/N* graph for a steel, showing effect of carburisation. Note: number of cycles shown on logarithmic scale

led to a fatigue failure. *Figure 1.30* shows the effect on the fatigue properties of a steel of a small hole acting as a stress raiser. With the hole, at every stress amplitude value less cycles are needed to reach failure. There is also a lower fatigue limit with the hole present, 700 MN m^{-2} (MPa) instead of over 1000 MN m^{-2} (MPa).

Figure 1.32 shows the effect on the fatigue properties of a steel of exposure to salt solution. The effect of the corrosion resulting from the salt solution attack on the steel is to reduce the number of stress cycles needed to reach failure for every stress amplitude. The non-corroded steel has a fatigue limit of 450 MN m^{-2} (MPa), the corroded steel has no fatigue limit. There is thus no stress amplitude below which failure will not occur. The steel can be protected against the corrosion by plating; for example, chromium or zinc plating of the steel can result in the same *S/N* graph as the non-corroded steel even though it is subject to a corrosive atmosphere.

Residual stresses can be produced by many fabrication and finishing processes. If the stresses produced are such that the surfaces have compressive residual stresses then the fatigue properties are improved, but if tensile residual stresses are produced at the surfaces then poorer fatigue properties result. The case-hardening of steels by carburising results in compressive residual stresses at the surface, hence carburising improves the fatigue properties. *Figure 1.31* shows the effect of carburising a hardened steel. Many machining processes result in the production of surface tensile residual stresses and so result in poorer fatigue properties.

The effect of surface finish on the fatigue properties of a component is very significant. Scratches, dents or even surface identification markings can act as stress raisers and so reduce the fatigue properties. Shot peening a surface produces surface compressive residual stresses and improves the fatigue performance.

An increase in temperature can lead to a reduction in fatigue properties as a consequence of oxidation or corrosion of the metal surface increasing. For example, the nickel-chromium alloy

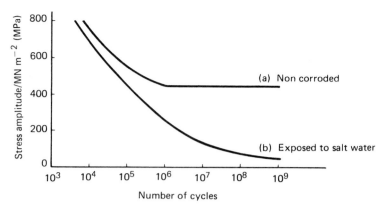

Figure 1.32 S/N graph for a steel (a) with no corrosion and (b) corroded by exposure to salt solution. Note: number of cycles shown on logarithmic scale

Nimonic 90 undergoes surface degradation at temperatures around 700 to 800°C and there is a poorer fatigue performance as a result. In many instances an increase in temperature does result in a poorer fatigue performance.

Example
The following information has been taken from data supplied by BCIRA for grey cast irons. What would be the expected fatigue limit for un-notched grey cast iron with a tensile strength of about (a) 200 N mm^{-2} (MPa), (b) 350 N mm^{-2} (MPa)?

'Fatigue limit
For grey cast irons with tensile strengths of 150–300 N mm^{-2} the fatigue limit is about 45 per cent of the tensile strength. At higher stresses the ratio decreases, and values of 0.425 and 0.38 have been considered typical of irons with tensile strengths of 350 and 400 N mm^{-2}.

Grey cast irons are relatively insensitive to notches in fatigue; normally a notch will not reduce the fatigue limit in an iron with a tensile strength of 150 N mm^{-2}. In an iron with a tensile strength of 400 N mm^{-2} notching has been found to reduce the fatigue limit to about 83 per cent of the unnotched value.'

(a) Fatigue limit is about 45% of 200 N mm^{-2}, i.e. 90 N mm^{-2} (MPa).

(b) Fatigue limit is about 42.5% of 350 N mm^{-2}, i.e. about 149 N mm^{-2} (MPa).

LONG-TERM BEHAVIOUR – CREEP

There are many situations where a piece of material is exposed to a stress for a protracted period of time. The stress/strain data obtained from the conventional tensile test refer generally to a situation where the stresses are applied for quite short intervals of time and so the strain results refer only to the immediate values resulting from stresses. Suppose stress were applied to a piece of material and the stress remained acting on the material for a long time – what would be the result? If you tried such an experiment with a strip of lead you would find that the strain would increase with time – the material would increase in length with time even though the stress

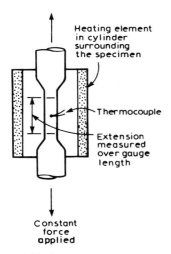

Figure 1.33 A creep test

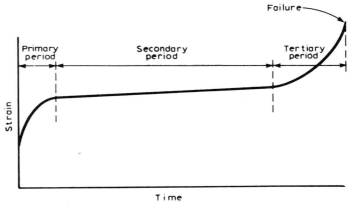

Figure 1.34 Typical creep curve for a metal

remained constant. This phenomenon is called *creep*, which can be defined as the continuing deformation of a material with the passage of time when the material is subject to a constant stress.

For metals, other than the very soft metals like lead, creep effects are negligible at ordinary temperatures, but however become significant at higher temperatures. For plastics, creep is often quite significant at ordinary temperatures and even more noticeable at higher temperatures.

Figure 1.33 shows the essential features of a creep test. A constant stress is applied to the specimen, sometimes by the simple method of suspending loads from it. Because creep tests with metals are usually performed at high temperatures a furnace surrounds the specimen, the temperature of the furnace being held constant by a thermostat. The temperature of the specimen is generally measured by a thermocouple attached to it.

Figure 1.34 shows the general form of results from a creep test. The curve generally has three parts. During the *primary creep* period the strain is changing but the rate at which it is changing with time decreases. During the *secondary creep* period the strain increases steadily with time at a constant rate. During the *tertiary creep* period the rate at which the strain is changing increases and eventually causes failure. Thus the initial stress, which did not produce early failure, will result in a failure after some period of time. Such an initial stress is referred to as the *stress to rupture* in some particular time.

Factors affecting creep behaviour of metals

For a particular material, the creep behaviour depends on both the temperature and the initial stress; the higher the temperature the greater the creep, also the higher the stress the greater the creep. *Figure 1.35* shows both these effects. Thus to minimise creep, the conditions need to be low stress and low temperature.

Figure 1.36 shows one way of presenting creep data, indicating the design stress that can be permitted at any temperature if the creep is to be kept within specified limits. In the example given, the limit is 1% creep in 10 000 hours. Thus for the Pireks 25/20 nickel-chrome alloy a stress of 58.6 N mm^{-2} (MPa) at a temperature of

24 Mechanical properties of metals

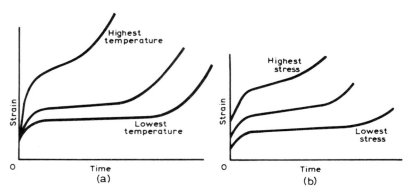

Figure 1.35 Creep behaviour of a material (a) at different temperatures but subject to constant stress, (b) at different stresses but subject to constant temperature

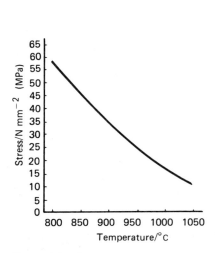

Figure 1.36 Data to give 1% creep in 10 000 h for Pireks 25/20 alloy (0.45% C, 0.8% Mn, 1.2% Si, 20% Cr, 25% Ni) (courtesy of Darwins Alloy Castings Ltd)

800°C will produce the 1% creep in 10 000 h. At 1050° a stress of only 10.3 N mm^{-2} (MPa) will produce the same creep in 10 000 h. *Figure 1.37* shows how stress to rupture the material in 10 000 h varies with temperature. At 800°C a stress of 65.0 N mm^{-2} will result in the Pireks 25/20 alloy failing in 10 000 h; at 1050°C a stress of only 14.5 N mm^{-2} (MPa) will result in failure in the same time.

Another factor that determines the creep behaviour of a metal is its composition. *Figure 1.38* shows how the stress to rupture different materials in 1000 h varies with temperature. Aluminium alloys fail at quite low stresses when the temperature rises over 200°C. Titanium alloys can be used at higher temperatures before the stress to rupture drops to very low values, while stainless steel is even better and nickel-chromium alloys offer yet better resistance to creep.

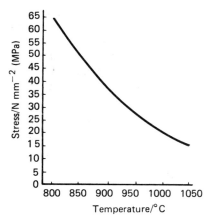

Figure 1.37 Data showing stress to rupture at 10 000 h for Pireks 25/20 alloy (see also *Figure 1.36*)

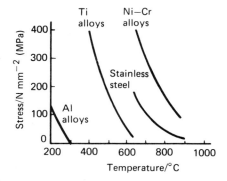

Figure 1.38 Stress to rupture in 1000 h for different materials

WHICH MECHANICAL PROPERTY?

Which mechanical properties need to be considered when choosing a material for a particular application? The following are some of the factors that need to be considered.

1 The nature of the loading

(a) Is the loading either just tension, compression or twisting, and is it applied slowly? This determines whether the data from tensile testing, compression testing or torsion testing can be considered.

(b) Is the loading shock loading? This determines whether impact tests give the most relevant data.

(c) Is the loading alternating? This determines whether fatigue testing is needed to give the relevant data. The form of the loading, e.g. twisting back-and-forth or bending back-and-forth, needs to be known so the relevant form of fatigue tests can be used.

2 The range and effects of loading

(a) What is the range of the stresses likely to be encountered in service?

(b) What elastic deformations are permissible?

For, say, a tensile, slowly applied load, these points give the minimum required values for the yield stress and strain to be specified for the material. Hence these can put restrictions on the materials that can be used.

(c) Is stiffness important?

For some components the most important requirement is that it is stiff. This means that a high tensile modulus material is required.

(d) Is strength per unit weight important?

For some applications an important requirement is that the material should have a high strength for its weight. If so, information on the tensile strength and the density of the material is required.

3 Environmental effects

(a) Is the environment a corrosive one? This will dictate the choice of material as that most able to resist the corrosive effects.

(b) Are high temperatures likely to be encountered? High temperatures could mean that creep becomes a major problem.

(c) Are low temperatures likely to be encountered? This could mean that a steel changes from being ductile to become brittle. Impact tests could reveal this.

The above are just some of the questions that need to be posed in considering the choice of a material and what data are required in order to make the choice. Other factors which can be important are:

Resistance to corrosion or degradation
Wear resistance and frictional properties
Thermal properties
Electrical properties
Magnetic properties
The possible methods for manufacture
Cost

Example

Consider the nature of the loading likely to be experienced by the following components and hence state what are likely to be the crucial mechanical properties for the materials used for the components.

(a) The legs supporting a North Sea oil rig.

(b) The casting on which the landing wheels of an aircraft are mounted.

(c) The metal tubing used for the aerial for a car radio.

(a) The legs will be subject to a steady loading due to the weight of the rig, have to operate at low temperatures and be subject to continuous buffeting by the sea. This would suggest that the material used will require good tensile properties. Since such properties would be required at low temperatures, an impact test could indicate whether the material would be brittle or ductile at such temperatures. The buffeting of the sea could lead to fatigue problems and so fatigue tests would be required.

(b) This will be subject to shock loading when the aircraft lands and thus an impact test would be required.

(c) This will flex back and forth when the car is being driven and so fatigue strength will be required.

PROBLEMS

1 Explain the terms: elasticity, plasticity, toughness, ductility, malleability.

2 Explain the terms: tensile strength, tensile modulus, limit of proportionality, yield stress, proof stress, percentage elongation, percentage reduction in area.

3 Explain why tensile test pieces have a standard relationship between gauge length and cross-sectional area.

4 Sketch the form of the stress–strain graphs for (a) brittle stiff materials, (b) brittle non-stiff materials, (c) ductile stiff materials, (d) ductile non-stiff materials.

5 The effect of working an aluminium alloy (1.25% manganese) is to change the tensile strength from 110 N mm^{-2} to 180 N m^{-2} and the percentage elongation from 30% to 3%. Is the effect of the working to (a) make the material stronger, (b) make the material more ductile?

6 A sand-cast aluminium alloy (12% silicon) is found to have a percentage elongation of 5%. Would you expect this material to be brittle or ductile?

7 An annealed titanium alloy has a tensile strength of 880 N mm^{-2} and a percentage elongation of 16%. A nickel alloy, also in the annealed condition, has a tensile strength of 700 N mm^{-2} and a percentage elongation of 35%. Which alloy is (a) the stronger, (b) the more ductile in the annealed condition?

8 Explain why the results of a tensile test can be influenced by (a) the position in a casting from which the test piece is taken, (b) the direction of the axis of the test piece when taken from a wrought material?

9 Explain why the tensile properties of carbon steels are quoted with reference to the section thickness of the material.

10 Describe the Izod or the Charpy impact test.

11 Describe the difference between the appearance of a brittle and a ductile fracture in an impact test piece.

12 Explain how impact tests can be used to determine whether a heat treatment process has been carried out successfully.

13 The following are Izod impact energies at different temperatures for samples of annealed cartridge brass (70% copper − 30% zinc). What can be deduced from the results?

Temperature/°C	+27	−78	−197
Impact energy/J	88	92	108

14 The following are Charpy V-notch impact energies for annealed titanium at different temperatures. What can be deduced from the results?

Temperature/°C	+27	−78	−196
Impact energy/J	24	19	15

15 Describe the principles of the Brinell, Vickers and Rockwell hardness measurement methods.

16 How can a hardness measurement give an indication of the tensile strength of a material?

17 Outline the limitations of the Brinell hardness test.

18 With Rockwell test results, a letter A, B, C, etc. is always given with the results. What is the significance of the letter?

19 Which hardness test could be used with thin steel sheet?

20 Explain what is meant by the Moh scale of hardness.

21 A sample of brass can just be scratched by calcite but not by gypsum. What would be its Moh hardness number?

22 Specify the type of test that could be used in the following instances.

(a) A large casting is to be produced and a check is required as to whether the correct cooling rate occurs.

(b) The storekeeper has mixed up two batches of steel, one batch having been surface hardened and the other not. How could the two be distinguished?

(c) What test could be used to check whether tempering has been correctly carried out for a steel?

(d) What test could be used to determine whether a metal has been correctly heat treated?

23 *Figure 1.39* shows the S/N graph for an aluminium alloy.

(a) For how many stress cycles could a stress amplitude of 140 MN m^{-2} be sustained before failure occurs?

(b) What would be the maximum stress amplitude that should be applied if the component made of the material is to last for 50 million stress cycles?

(c) The alloy has a tensile strength of 400 MN m^{-2} and a yield stress of 280 MN m^{-2}. What should be the limiting stress when such an alloy is used for static conditions? What should be the limiting stress when the alloy is used for dynamic conditions where the number of cycles is not likely to exceed 10 million?

24 *Figure 1.40* shows the stress rupture properties of two alloys, one 50% chromium and 50% nickel, the other (IN 657) 48–52% chromium, 1.4–1.7% niobium, 0.1% carbon, 0.16% nitrogen, 0.2% carbon + nitrogen, 0.5% maximum silicon, 1.0% iron, 0.3% maximum manganese and the remainder nickel. The creep rupture data is presented for two different times, 1000 h and 10 000 h.

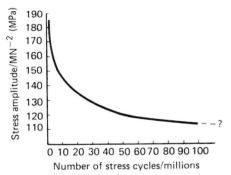

Figure 1.39 S/N graph for an aluminium alloy

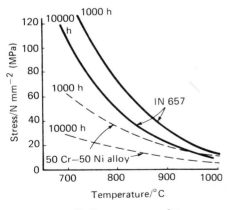

Figure 1.40 Creep rupture data from 'high chromium Cr–Ni alloys to resist residual fuel oil ash corrosion' (courtesy of Inco Europe Ltd)

(a) What is the significance of the difference between the 1000 h and 10 000 h graphs?

(b) What is the difference in behaviour of the 50 Cr–50 Ni alloy and the IN 657 alloy when temperatures increase?

(c) The IN 657 alloy is said to show 'improved hot strength' when compared with the 50 Cr–50 Ni alloy. Explain this statement.

25 State the nature of the loading likely to be experienced by the following components and hence what are likely to be the crucial mechanical properties for the materials used for the components.

(a) The leaf spring of a car suspension.
(b) The metal legs of a table.
(c) The drive shaft of a motor.
(d) A storage tank for liquid nitrogen.
(e) Tubes carrying hot gases.

2 Mechanical properties of polymeric materials

Objectives: At the end of this chapter you should be able to:
Interpret the results of tensile testing, hardness measurements, impact testing, fatigue testing and creep measurements.
Describe the effects of temperature and age on the properties.
Select the relevant property data for a given service requirement for a component.

POLYMERIC MATERIALS

A polymer is a very large molecule comprising hundreds or even thousands of atoms. Such molecules are formed by linking together one or two types of small molecules to form chain or network structures. Polymers can be grouped into three categories: thermoplastics, thermosets and elastomers. Thermoplastics are based on chain structures. When heated, such materials soften, become plastic, and on cooling reharden. Thermosets are based on network structures. When heated they do not soften but char and decompose. Elastomers have structures, based on chains, which allow considerable extensions which are reversible – think of the rubber band.

The term 'plastic' is used for polymeric materials, other than elastomers, to which other materials have been added in order to produce specific properties (*see* Chapter 6 for more details).

MECHANICAL PROPERTIES

The main terms used in describing the mechanical properties of polymeric materials are *elasticity*, *plasticity*, *strength*, *toughness*, *fatigue strength*, *ductility*, *brittleness* and *hardness*. The terms are defined in exactly the same way as for metals, *see* Chapter 1.

The main general ways in which the properties of polymeric materials differ from those of metals are:

1 Lower density.
2 Less stiffness.
3 Lower strength.
4 Less hardness.
5 Their mechanical properties may deteriorate rapidly with quite small increases in temperature.
6 Their mechanical properties may change with time; creep is a greater problem.
7 Their mechanical properties can be affected by exposure to ultraviolet radiation in sunlight.
8 High electrical resistivity, being electrical insulators whereas metals are good electrical conductors.
9 Bad conductors of heat, unlike metals which are good conductors.
10 Higher expansivity.

THE TENSILE TEST

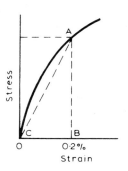

Figure 2.1 Part of the stress–strain graph for a plastic. The secant modulus is AB/BC, i.e. the slope of the line AC

Tensile tests can be used with polymeric test pieces to obtain stress–strain data. The term 'tensile strength' has the same meaning as with metals and the tensile modulus can be determined in the same way as with metals, i.e. the slope of the stress–strain graph for stresses below the limit of proportionality. For many polymeric materials there is, however, no straight-line part of the stress–strain graph and thus the tensile modulus cannot be determined in the way specified for metals. In such cases it is common practice to quote a modulus, termed the *secant modulus*, which is obtained by dividing the stress at a value of 0.2% strain by that strain (*Figure 2.1*).

Polymeric materials have relatively low values of tensile or secant modulus when compared with metals, e.g. polythene has a modulus of about 0.1 to 1.2 kN mm^{-2} (GPa) compared with values of the order of 100 kN mm^{-2} (GPa) for metals. Hence polymeric materials are considerably less stiff than metals.

Polymeric materials have tensile strengths lower than those of metals, e.g. polythene has a tensile strength of between 4 and 38 N mm^{-2} (MPa) compared with values of the order of a few hundred to more than a thousand N mm^{-2} (MPa) for metals.

The tensile modulus and the tensile strength of polymeric materials can however be readily modified by incorporating fillers in the material (*see* Chapter 6 for more details).

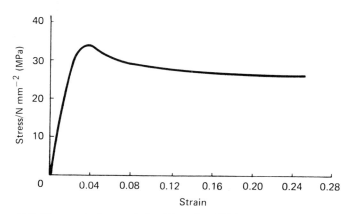

Figure 2.2 Stress–strain graph for Novodur PK (courtesy of Bayer UK Ltd)

Example
Figure 2.2 shows the stress–strain graph for a sample of ABS Novodur grade PK (courtesy of Bayer (UK) Ltd). Estimate (a) the tensile modulus and (b) the tensile strength.

(a) The tensile modulus is the slope of the stress–strain graph below the limit of proportionality, about 1.4 kN mm^{-2} (GPa) in this case.

(b) The tensile strength is the maximum stress, about **34 N mm^{-2}** (MPa) in this case.

Ductile and brittle polymeric materials

Thermosetting plastics tend to behave as brittle materials. Thermoplastic materials can be either brittle or ductile depending on the temperature, as can also be elastomers. *Figure 2.3* shows the typical forms of stress–strain graphs for brittle and ductile polymeric materials.

Melamines are thermosetting materials and have percentage elongations (*see* Chapter 1 for a definition) of about 1% or less. Melamines are brittle materials. High density polythene, a thermoplastic, can have percentage elongations as high as 800%. It is ductile.

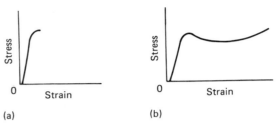

Figure 2.3 Stress–strain graphs for (a) a brittle polymeric material, (b) a ductile polymeric material

Example
The percentage elongation of a thermoplastic was found to be 40% immediately after it had been used to produce a moulding and was dry. This later changed to 160% when the material had been standing in air for some time and had absorbed water vapour. What has been the effect on the properties of the material of this absorption of water vapour?

The increase in the percentage elongation indicates an increase in ductility of the material.

Validity of tensile test data

The stress–strain properties of a particular polymeric material depend on:
1 The strain rate
2 Temperature
3 Degree of crystallinity of the material
4 Additives
5 Orientation
6 Time

The stress–strain properties of polymeric materials are much more dependent on the rate at which strain is applied than are those of metals. For example, the stress–strain data may indicate a yield stress of 62 N mm^{-2} (MPa) when the rate of elongation is 12.5 mm/min, but 74 N mm^{-2} (MPa) when it is 50 mm/min. *Figure 2.4* shows the general forms of stress–strain graphs for polymeric materials at different strain rates. Note that at high strain rates a material that was ductile at lower strain rates may become brittle.

Another factor that has considerable effect on the stress–strain properties of polymeric materials is temperature. For example, polyacetal, a crystalline thermoplastic, has a yield stress of 110 N mm^{-2} (MPa) at $-40°C$, 60 N mm^{-2} (MPa) at $20°C$ and

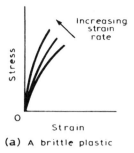 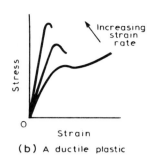

(a) A brittle plastic   (b) A ductile plastic

Figure 2.4 The effect of strain rate on the stress–strain graphs for plastics

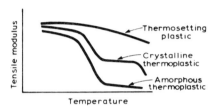

Figure 2.5 The effect of temperature on the tensile modulus of plastics

45 N mm^{-2} (MPa) at 80°C. *Figure 2.5* shows the general trend in tensile modulus with temperature for plastics. An increase in temperature produces quite noticeable changes in yield stress and tensile modulus for a thermoplastic.

For those polymeric materials that can have crystalline structures, the degree of crystallinity can markedly affect the stress–strain properties. Thus polythene with 60% crystallinity has a tensile strength of about 8 to 16 N mm^{-2} (MPa) and a tensile modulus of 0.1 to 0.3 kN mm^{-2} (GPa), while the same material with 95% crystallinity has a tensile strength of 22 to 38 N mm^{-2} (MPa) and a tensile modulus of 0.4 to 1.3 kN mm^{-2} (GPa). The more crystalline material is stronger and stiffer.

Plastics are generally a mixture of a polymer and various other substances. These can have a significant effect on the stress–strain properties. For example, a polyester has a tensile modulus of 2 to 4 kN mm^{-2} (GPa) and a tensile strength of 20 to 70 N mm^{-2} (MPa). The inclusion of short glass fibres in the polyester can result in a tensile modulus of the order of 10 kN mm^{-2} (GPa) and a tensile strength of 110 N mm^{-2} (MPa). The material is stiffer and stronger.

The stress–strain properties of plastics can show considerable variation with direction as a result of the orientation of the polymer molecules during the production process, e.g. in extrusion. A plastic that includes fibres aligned in some particular direction can show quite marked differences in different directions. For instance, the tensile strength along the direction of the fibres might be of the order of 500 N mm^{-2} (MPa) while at right angles to the aligned fibres the tensile strength it was only 30 N mm^{-2} (MPa).

The stress–strain relationships obtained as a result of a tensile test at a constant rate of elongation can be useful in comparing one material with another or for quality control, but may be misleading when used to predict the behaviour of the material in service. This is because the properties of polymeric materials change with time. A polymeric material under load can continue to change shape for the length of time for which the load is applied. A load which, when initially applied, was insufficient to cause yielding of the material might, with time, be able to do this. The tensile test data is immediate data without any attempt to take account of time. For this reason, stress–strain graphs are often derived from creep data in

Mechanical properties of polymeric materials 33

order to take account of time, and are then known as isochronous stress–strain graphs. These and other graphs are looked at in more detail in the next section of this chapter.

CREEP BEHAVIOUR OF PLASTICS

For metals, creep is significant mainly at high temperatures, but it can be significant for plastics at normal temperatures. The creep behaviour of a plastic depends on temperature and stress, just as with metals. It also depends on the type of plastic involved – flexible plastics show more creep than stiff ones.

Figure 2.6 shows how the strain on a sample of polyacetal at 20°C varies with time for different stresses. The higher the stress, the greater the creep. As can be seen from the graph, the plastic creeps quite substantially in a period of just over a week, even at relatively low stresses.

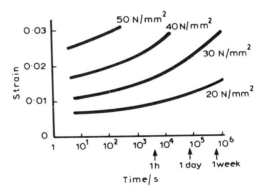

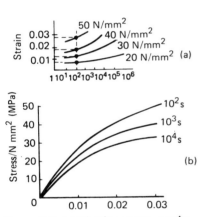

Figure 2.6 Creep behaviour of polyacetal at different stresses. Note logarithmic time scale

Figure 2.7 (a) Plotting stress–strain data, (b) Isochronous stress–strain graph

Figure 2.6 is the form of graph obtained by plotting results derived from creep tests. On the basis of this graph of strain against time at different stresses, a graph of stress against strain for different times can be produced. Thus for a time of 10^2 s, a vertical line drawn on *Figure 2.6* enables the stresses producing different strains after this time to be read from the graph (*Figure 2.7a*). The resulting stress–strain graph, shown in *Figure 2.7b*, is known as an *isochronous stress–strain graph*. For a specified time, the quantity obtained by dividing the stress by the strain for the isochronous stress–strain graph can be calculated; this value is known as the *creep modulus*. It is not the same as Young's modulus, though it can be used to compare the stiffness of plastics. The creep modulus varies with both time and strain; *Figure 2.8* shows how, at 0.5% strain and 20°C, it varies with time for the polyacetal described in *Figures 2.6* and *2.7*.

Figure 2.9 shows how the stress to rupture a plastic – Durethan, a polyamide – varies with time at different temperatures. The higher the temperature, the lower the stress needed to rupture the material after a given time.

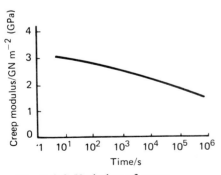

Figure 2.8 Variation of creep modulus with time for polyacetal at 0.5% strain and 20°C

HARDNESS MEASUREMENTS

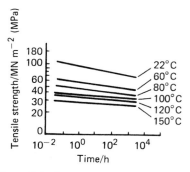

Figure 2.9 Creep rupture graph for Durethan BKV 30 (courtesy of Bayer UK Ltd)

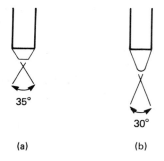

Figure 2.10 Indenters with Shore durometer test

The Brinell, Vickers and Rockwell tests can be used with polymeric materials. The Rockwell test, measuring penetration depth rather than the surface area of indentation is more widely used. Scales M and R are generally used (*see* Chapter 1 for more details), the M scale being used for the harder polymeric materials and the R scale for softer materials. The following are some typical values.

Thermoplastics
PVC, no plasticiser	110 HRR
Polystyrene	80 HRM
ABS	70 HRM
Nylon 6	110 HRR

Thermosets
Melamine formaldehyde	120 HRM
Phenol formaldehyde	115 HRM
Urea formaldehyde	115 HRM
Polyester, unfilled	117 HRM

Another form of test is the *Shore durometer*. Two scales are used with polymeric materials. The type A scale is obtained by pressing an indenter in the form of a truncated cone (*Figure 2.10a*) into the surface with a force of 8 N. The depth of indentation immediately after this is done is the measure of the hardness. The type A scale is used for soft polymeric materials. The type D scale is used with harder polymeric materials and involves a spherically-ended cone (*Figure 2.10b*) being pressed into the surface under a load of 44.5 N. The following are some typical values.

Elastomers
Natural rubber	30–90 A
Neoprene	30–90 A
Butyl	40–80 A
Silicone	30–85 A

An alternative to hardness is the British Standard test (BS 2782) which gives a *softness number*. An indenter, a ball of diameter 2.38 mm, is pressed against the polymeric material by an initial force of 0.294 N for 5 s. Then an additional force of 5.25 N is applied for 30 s. The difference between the two penetration depths is measured and expressed as the softness number. This is just the depth expressed in units of 0.01 mm. Thus a penetration depth of 0.05 mm is a softness number of 5. The test is carried out at a temperature of $23 \pm 1 °C$.

IMPACT TESTS

The Charpy and Izod impact tests are used with polymeric materials (*see* Chapter 1 for details of the tests). The results for polymeric materials are, however, often given as absorbed energy divided by either the cross-sectional area of the un-notched test piece or the cross-sectional area behind the notch in the case of notched test pieces. The impact strength depends on the notch-tip radius used and results are often quoted for a range of notch-tip radii. The impact strength also depends on the temperature.

The following are some typical impact strength values for a notch-tip radius of 0.25 mm with a notch depth of 2.75 mm, used at about room temperature.

Material	Impact strength/kJ m^{-2}
Polythene, high density	30
Nylon 6.6	5
PVC, un-plasticised	3
Polystyrene	2
ABS	25

FATIGUE PROPERTIES OF PLASTICS

Fatigue tests can be carried out on plastics in the same way as with metals. A factor not present with metals is that when a plastic is subject to an alternating stress it becomes significantly warmer. The faster the stress is alternated, i.e., the higher the frequency of the alternating stress, the greater the temperature rise. Under very high frequency alternating stresses, the temperature rise may be large enough to melt the plastic. To avoid this, fatigue tests are normally carried out with alternating stresses at lower frequencies than are used with metals. The results of such tests, however, are not entirely valid if the alternating stresses experienced by the plastic component in service are higher than those used for the test.

Figure 2.11 shows an *S/N* graph for a plastic, unplasticised PVC. The alternating stresses were applied with a square waveform at a frequency of 0.5 Hz, i.e. a change of stress every 2 s. The graph seems to indicate that there will be no stress amplitude for which failure will not occur; the material thus seems to have no fatigue limit.

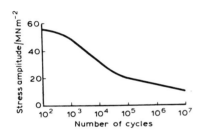

Figure 2.11 *S/N* graph for unplasticised PVC, alternating stress being a square waveform at frequency 0.5 Hz. Note: number of cycles shown on logarithmic scale

SERVICE REQUIREMENTS

In selecting a polymeric material for a component the following points merit consideration.

1 Does it have to be stiff? Such a requirement would mean a material with a high tensile modulus. This might suggest a fibre reinforced polymeric material.

2 Does it have to be strong? Such a requirement would indicate a high tensile modulus. This might suggest a fibre reinforced polymeric material.

3 Does it have to be tough? This means a material which can withstand impact abuse and so one with a high impact strength.

4 Does it have to withstand loads for long periods of time? Such a requirement would mean a material which was not very susceptible to creep.

5 Does it have to maintain its properties at high temperatures (perhaps 100°C)? Many thermoplastics deteriorate in their properties quite markedly when the temperature rises; thermosets deteriorate less. There are, for example, few polymeric materials that can be used for domestic hot water pipes.

6 Does it have to be resistant to chemical attack? Plastics, for example, are generally resistant to weak acids and weak alkalis but

may be attacked by strong acids which cause discoloration and embrittlement. Some liquids may cause swelling and softening; water does this with many plastics.

7 Does it have to withstand sunlight without degradation of properties? Many polymeric materials are damaged by exposure to the ultraviolet rays in sunlight. The effects can be minimised by suitable additives incorporated in the material.

8 Does it have to be in use over a long period? Most plastics and rubbers show a deterioration of properties with time as a result of exposure to oxygen in the air, particularly at higher temperatures. This effect is called *ageing*.

9 Are special properties required, e.g. optical transparency, electrical and thermal properties?

10 Are there any limitations on processing methods? Products required in continuous lengths are most likely to be extruded. Most thermoplastics can be extruded.

11 What limits are there on cost? Some materials are more expensive than others, some processing methods are more expensive than others.

PROBLEMS

1 *Figure 2.12* shows part of the stress–strain graph for a sample of nylon 6. Estimate (a) the tensile modulus and (b) the tensile strength for the sample.

2 What is the secant modulus of elasticity?

3 What is the effect of strain rate on the data obtained from tensile tests for plastics?

4 Cellulose acetate has a tensile modulus of $1.5 \, \text{kN mm}^{-2}$ while polythene has a tensile modulus of $0.6 \, \text{kN mm}^{-2}$. Which of the two plastics will be the stiffer?

5 A plastic is modified by the inclusion of glass fibres. What test can be used to determine whether this has made the plastic stiffer?

6 *Figure 2.13* shows the stress–strain graphs for two grades of Makrolon and *Figure 2.14* their variation of tensile modulus with temperature.

(a) On the basis of these graphs, which of the grades will be at room temperature (i) the stiffest, (ii) the strongest, (iii) the most ductile?

(b) What is the limiting temperature at which these materials can be used and remain stiff?

7 Explain how an isochronous stress/strain graph for a polymer can be obtained from creep test results.

8 Explain the significance of the graph shown in *Figure 2.9* for the creep rupture behaviours of Durethan.

9 Durethan, as described by *Figure 2.9*, is used for car fan blades, fuse box covers, door handles and plastic seats. How would the behaviour of the material change with increased temperature or stress?

10 *Figure 2.15* shows how the strain changes with time for two different polymers when they are subjected to a constant stress. Describe how the materials will creep with time. Which material will creep the most?

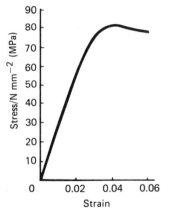

Figure 2.12 Part of the stress–strain graph for nylon 6 (Durethan SK, Bayer UK Ltd)

Mechanical properties of polymeric materials

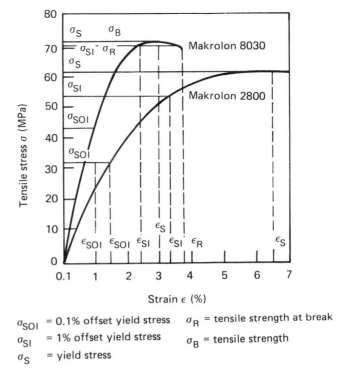

σ_{SOI} = 0.1% offset yield stress σ_R = tensile strength at break
σ_{SI} = 1% offset yield stress σ_B = tensile strength
σ_S = yield stress

Figure 2.13 Stress–strain diagram of tensile strength test to DIN 53455 using Makrolon 2800 and 8030 (courtesy of Bayer UK Ltd)

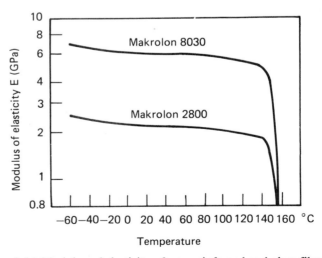

Figure 2.14 Modulus of elasticity of non-reinforced and glass fibre reinforced Makrolon as a function of temperature (courtesy of Bayer UK Ltd)

11 What tests are commonly used to determine the hardness of polymeric materials?

12 The following are Charpy impact strengths for nylon 6.6 at different temperatures. What can be deduced from the results?

Temperature/°C	−23	−33	−43	−63
Impact strength/kJ m^{-2}	24	13	11	8

13 The impact strength of samples of nylon 6, at a temperature of 22°C, is found to be 3 kJ m^{-2} in the 'as moulded' condition, but 25 kJ m^{-2} when the sample has gained 2.5% in weight through water absorption. What can be deduced from the results?

14 Makrolon (*see* Problem 6) is used to make cups and saucers. The manufacturer (Bayer UK Ltd) states in the material specification that 'although tableware made of Makrolon, for example, can be washed innumerable times in hot water, the material cannot be used continuously in water at temperatures over approximately 60°C without restrictions. Hot water produces gradual chemical deterioration accompanied by a reduction in impact strength?

   (a) What will be the effect of a reduction in impact strength?
   (b) How could this change be measured?

15 Plot the *S/N* graph for the plastic (cast acrylic) which gave the following fatigue test results when tested with a square waveform at 0.5 Hz. Why specify this frequency? What is the endurance limit at 10^6 cycles?

Stress amplitude/MM m^{-2} (MPa)	Number of cycles before failure
70	10^2
62	10^3
58	10^4
55	10^5
41	10^6
31	10^7

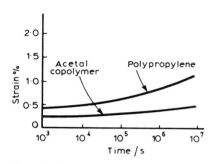

Figure 2.15

16 What is meant by ageing when applied to polymeric materials?

17 What in general is the effect of an increase in temperature on the properties of polymeric materials?

3 Alloying of metals

Objectives: At the end of this chapter you should be able to:

Explain what is meant by the term 'alloy' and give reasons why alloys are generally used in engineering in preference to pure metals.

Distinguish between mixtures, solutions and compounds.

Explain the term 'phase'.

Explain how thermal equilibrium diagrams are produced and interpret such diagrams.

Deduce the structures forming on the solidification of alloys, given the thermal equilibrium diagram.

ALLOYS Brass is an alloy composed of copper and zinc. Bronze is an alloy of copper and tin. An *alloy* is a metallic material consisting of an intimate association of two or more elements. The everyday metallic objects around you will be made almost invariably from alloys rather than the pure metals themselves. Pure metals do not always have the appropriate combination of properties needed; alloys can, however, be designed to have them.

The coins in your pocket are made of alloys. Coins need to be made of a relatively hard material which does not wear away rapidly, i.e. the coins have to have a 'life' of many years. Coins made of, say, pure copper would be very soft; not only would they suffer considerable wear but they would bend in your pocket.

Coins (British)	Percentage by mass			
	Copper	Tin	Zinc	Nickel
1p, 2p	97	0.5	2.5	—
5p, 10p, 50p	75	—	—	25

Pure metals tend to be soft and to have high ductility, low tensile strength and low yield strength. Because of this they are rarely used in engineering. Alloying can produce harder materials with higher tensile strength and higher yield stress, and with a reduction in ductility. Such materials are more useful in engineering. There are, however, some circumstances in which the properties of pure metals are useful. These are where high electrical conductivity is required (alloying reduces conductivity); where good corrosion resistance is required (alloying can result in less corrosion resistance); and where very high ductility is required.

IRON ALLOYS Pure iron is a relatively soft material and is of hardly any commercial use in that state. Alloys of iron with carbon are, however, very widely used. The following table indicates the names given to general groups of such alloys.

Material	Percentage carbon
Wrought iron	0 to 0.05
Steel	0.05 to 2
Cast iron	2 to 4.3

The percentage of carbon alloyed with iron has a profound effect on the properties of the alloy. The term *carbon steel* is used for those steels in which essentially just iron and carbon are present. The term *alloy steel* is used where other elements are included.

Carbon steels are grouped according to their carbon content. *Mild steel* is a group of steels having between 0.10% and 0.25% carbon, *medium-carbon steel* has between 0.20% and 0.50% carbon, *high-carbon steel* has more than 0.5% carbon. *Figure 3.1* shows how the mechanical properties of carbon steels depend on the percentage of carbon. Increasing the percentage of carbon:

1 Increases the tensile strength
2 Increases the hardness
3 Decreases the percentage elongation
4 Decreases the Charpy impact strength

Mild steel is a general purpose steel and is used where hardness and tensile strength are not the most important requirements. Typical applications are sections for use as joists in buildings, bodywork for cars and ships, screws, nails, wire. Medium carbon steel is used for agricultural tools, fasteners, dynamo and motor shafts, crankshafts, connecting rods, gears. High carbon steel is used for withstanding wear, where hardness is a more necessary requirement than ductility. It is used for machine tools, saws, hammers, cold chisels, punches, axes, dies, taps, drills, razors. The main use of high carbon steel is thus as a tool steel.

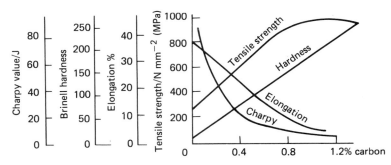

Figure 3.1 Properties of carbon steels

Example
Carbon steel is used for car bumpers. Which type of steel might be suitable?

A car bumper needs to be reasonably ductile so that it will 'give' and so absorb the energy of a collision. It does however have to be reasonably hard so that it is not too easily marked. These arguments would tend to suggest a medium-carbon steel. Another line of argument might be to suggest that car bumpers need to have high strength and hardness so that they can withstand collisions and so prevent damage to the weaker bodywork behind it. This would suggest the need for a high-carbon steel.

Car bumpers are generally made from a high-carbon steel having about 0.8% carbon, though steel with as little as 0.1% carbon has been used.

Example
Carbon steel is used for the heads of pick-axes. Which type of steel might be suitable?

The pick-axe head will need to be very hard and possess high strength. This would suggest the use of a high-carbon steel.

Pick-axe heads are generally made from a high-carbon steel having about 1% carbon.

COPPER ALLOYS

Pure copper is a soft material with low tensile strength. For many engineering purposes it is alloyed with other metals. The exception is where high electrical conductivity is required. Pure copper has a better conductivity than the alloys. The following indicates the names given to the various types of copper alloy.

Constituent metals	*Alloy name*
copper and zinc	brasses
copper and tin	bronzes
copper and tin and phosphorus	phosphor bronzes
copper and tin and zinc	gunmetals
copper and aluminium	aluminium bronzes
copper and nickel	cupro-nickels
copper and zinc and nickel	nickel silvers
copper and silicon	silicon bronze
copper and beryllium	beryllium bronze

Figure 3.2 shows how the percentage of zinc included with the copper in brasses affects the mechanical properties. Brasses with between 5 and 20% zinc are called gilding metals and, as the name implies, are used for architectural and decorative items to give a 'gilded' or golden colour. Cartridge brass is brass with 30% zinc;

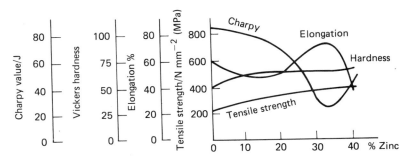

Figure 3.2 Properties of brasses

one of its main uses is for cartridge cases, items which require high ductility for the deep drawing process. The term 'basis brass' is used for the brass with 37% zinc. This is a good general cold-working alloy and is used for fasteners and electrical connectors. It does not have the high ductility of those brasses with less zinc. Brass with 40% zinc is called Muntz metal.

MIXTURES, SOLUTIONS AND COMPOUNDS

If you put sand in water, the sand does not react with the water but retains its identity, as does the water. The sand and water are said to form a mixture. In a *mixture*, each component retains its own physical structure and properties.

Sodium is a very reactive substance, which has to be stored under oil to stop it interacting with the oxygen in the air, and chlorine is a poisonous gas. Yet when these two substances interact, the product, sodium chloride, is eaten by you and me every day. The product is common salt. Sodium chloride is a compound. In a *compound* the components have interacted and the product has properties different from those of its constituents.

If you drop a pinch of common salt, sodium chloride, into cold water it will dissolve. The sodium chloride is said to be *soluble* in the cold water and the result is called a *solution*. In the solution the salt and the water retain their separate identities but the mixture is completely homogeneous throughout and separation does not occur over time. However, if sand is mixed with water, though initially the sand may be dispersed evenly, with the passage of time it will separate from the water. The sand is said to be *insoluble* in water.

The terms 'soluble' and 'insoluble' can also be used when we mix two liquids. Oil and water are insoluble in each other; when the two are mixed the result is just a mixture with the oil and the water retaining their separate identities. When alcohol is mixed with water the result is a solution; the water and the alcohol are said to be soluble in each other. It is not possible to identify the water and the alcohol as separate entities in the solution. When two liquid metals are mixed, say copper and nickel, a solution is produced in that it is not possible to identify in the liquid either the copper or the nickel. The copper and the nickel are said to be soluble in each other in the liquid state. Cadmium and bismuth are soluble in each other in the liquid state but insoluble in each other in the solid state. Copper and nickel, however, are soluble in each other in the solid state. The resulting copper–nickel alloy has, in the solid state, a structure in which it is impossible to distinguish the copper from the nickel. The copper and nickel are thus said to form a *solid solution*.

SOLUBILITY AND PRECIPITATION

Up to 36 g of sodium chloride can be dissolved in 100 g of cold water; more than that amount will not dissolve. With the 36 g dissolved in 100 g the resulting solution is said to be *saturated*. The *solubility* of sodium chloride in cold water is said to be 36 g per 100 g of water. If 40 g of sodium chloride is put into 100 g of cold water, only 36 g of it will dissolve, the remaining 4 g remaining as solid.

Alloying of metals 43

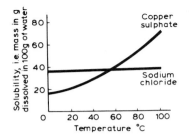

Figure 3.3 The solubility variation with temperature for sodium chloride, common salt, and copper sulphate. The solubility of copper sulphate increases considerably more with temperature than does that of sodium chloride

The solubility of sodium chloride in water depends on the temperature, hot water dissolving more sodium chloride than cold water. The solubility of sodium chloride in water does not, however, vary to a considerable extent with temperature, only slightly increasing as the temperature increases (*Figure 3.3*).

The solubility of copper sulphate in water does, however, increase quite significantly with temperature, as *Figure 3.3* shows. *Figure 3.4* shows this solubility variation with temperature plotted in a different way, the temperature axis being vertical rather than horizontal. The reason for this will become apparent later in this chapter when phase diagrams are considered. Suppose we have 40 g of copper sulphate dissolved in 100 g of hot water, say at 90°C. At this temperature the solution will not be saturated. If now the temperature decreases, then at a temperature of just under 60°C the solution will become saturated. Further cooling will result in the excess copper sulphate coming out of solution as a precipitate. At 0°C there will only be about 17 g still in solution.

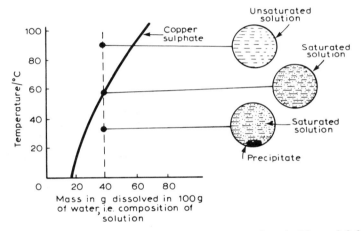

Figure 3.4 An alternative way of plotting the data given in Figure 3.3 for the solubility of copper sulphate in water

PHASE

A phase can be defined as a state of a material in which chemical composition and structure are uniform throughout. A piece of pure copper which throughout has the face-centred cubic structure, has but a single phase. Molten copper, at a higher temperature, represents a different phase of the same material, in that the arrangement of the atoms in liquid copper is different from that in the solid copper.

A completely homogeneous substance at a particular temperature has only one phase at that temperature. If you take any piece of that homogeneous substance it will show the same composition and structure.

Consider as an example the phase changes that occur when a common salt solution at room temperature is cooled. The solution is completely homogeneous throughout at room temperature, since the salt is soluble in water. Let us suppose that at room temperature the solution is saturated, i.e. the solution contains the maximum amount of salt that is possible. When such a solution is cooled, the solubility decreases and so salt will settle out of the solution. Thus we then have two phases, one phase being the saturated solution and the other phase the salt that has settled out. Further cooling results in freezing. When this begins, three phases are present: salt that has separated out; solution that has not frozen; and ice crystals. Further cooling will result in just two phases, salt and ice.

Liquid copper and liquid nickel are completely miscible, as are most liquid metals. This liquid copper–nickel solution is completely homogeneous and thus, at the temperature at which the two are liquid, there is but one phase present. When the liquid is cooled it begins to solidify. During this transition two phases are present, the liquid solution and solid metal. When it is all solid we have, in this particular case of copper–nickel, just one phase. This is because in the solid state the two metals are completely soluble in each other, the state being called a solid solution.

It is through an understanding of the phases present in alloys that their properties can be explained. Thus, for instance, the properties of carbon steels depend on the percentage of carbon present in the alloy (*Figure 3.1*) and this percentage determines the phases present at room temperature.

ALLOY TYPES

When an alloy is in a liquid state the atoms of the constituents are distributed at random through the liquid. When solidification occurs a number of possibilities exist.

1 The two components separate out with each in the solid state maintaining its own separate identity and structure. The two components are said to be *insoluble* in each other in the solid state.

2 The two components remain completely mixed in the solid state. The two components are said to be *soluble* in each other in the solid state, the components forming a solid solution.

3 On solidifying, the two components may show *limited solubility* in each other.

4 In solidifying, the elements may combine to form *intermetallic compounds*.

EQUILIBRIUM DIAGRAM

When pure water is cooled to 0°C it changes state from liquid to solid, i.e. ice is formed. *Figure 3.5* shows the type of graph that is produced if the temperature of the water is plotted against time during a temperature change from above 0°C to one below 0°C. Down to 0°C the water only exists in the liquid state. At 0°C solidification starts to occur and while solidification is occurring the temperature remains constant. Energy is still being extracted from the water but there is no change in temperature during this change of state. This energy is called *latent heat*. The *specific latent heat of fusion* is defined as the energy taken from, or given to, 1 kg of a substance when it changes from liquid to solid, or solid to liquid, without any change in temperature occurring.

Alloying of metals   45

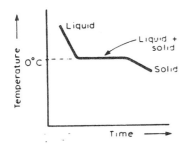

Figure 3.5 Cooling curve for water during solidification

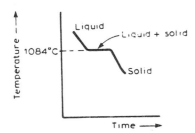

Figure 3.6 Cooling curve for copper during solidification

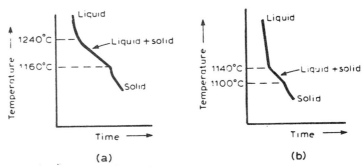

Figure 3.7 Cooling curves for copper–nickel alloys (a) 70% copper – 30% nickel, (b) 90% copper – 10% nickel

All pure substances show the same type of behaviour as the water when they change state. *Figure 3.6* shows the cooling graph for copper when it changes state from liquid to solid.

During the transition of a pure substance from liquid to solid, or vice versa, the liquid and solid are both in existence. Thus for the water, while the latent heat is being extracted there is both liquid and ice present. Only when all the latent heat has been extracted is there only ice. Similarly with the copper, during the transition from liquid to solid at 1084°C, while the latent heat is being extracted both liquid and solid exist together.

The cooling curves for an alloy do not show a constant temperature occurring during the change of state. *Figure 3.7* shows cooling curves for two copper–nickel alloys. With an alloy, the temperature is not constant during solidification. The temperature range over which this solidification occurs depends on the relative proportions of the elements in the alloy. If the cooling curves are obtained for the entire range of copper–nickel alloys a composite diagram can be produced which shows the effect the relative proportions of the constituents have upon the temperatures at which solidification starts and that at which it is complete. *Figure 3.8* shows such a diagram for copper–nickel alloys.

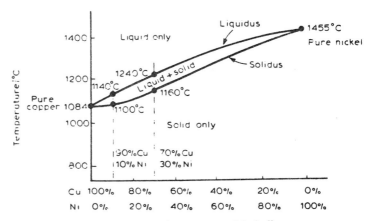

Figure 3.8 Equilibrium diagram for copper–nickel alloys

Thus for pure copper there is a single temperature point of 1084°C, indicating that the transition between liquid and solid takes place at a constant temperature. For 90% copper − 10% nickel the transition between liquid and solid starts at 1140°C and terminates at 1100°C when all the alloy is solid. For 70% copper − 30% nickel the transition between liquid and solid starts at 1240°C and terminates at 1160°C when all the alloy is solid.

The line drawn through the points at which each alloy in the group of alloys ceases to be in the liquid state and starts to solidify is called the *liquidus line*. The line drawn through the points at which each alloy in the group of alloys becomes completely solid is called the *solidus line*. These liquidus and solidus lines indicate the behaviour of each of the alloys in the group during solidification. The diagram in which these lines are shown is called the *thermal equilibrium diagram*.

The thermal equilibrium diagram is constructed from the results of a large number of experiments in which the cooling curves are determined for the whole range of alloys in the group. The diagram provides a forecast of the states that will be present when an alloy of a specific composition is heated or cooled to a specific temperature. The diagrams are obtained from cooling curves produced by very slow cooling of the alloys concerned. They are slow because time is required for equilibrium conditions to obtain at any particular temperature, hence the term 'thermal equilibrium diagram'.

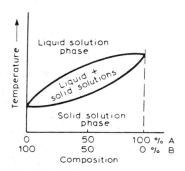

Figure 3.9 Equilibrium diagram for two metals that are completely soluble in each other in the liquid and solid states

Example

What phase is present at 1200°C for the 40% copper − 60% nickel alloy?

Using *Figure 3.8* the line for the 40% copper − 60% nickel alloy meets the solidus at a temperature greater than 1200°C. Hence the phase present at that temperature is the solid.

Example

A copper–nickel alloy is required to be liquid at temperatures down to 1140°C. What is the alloy with the highest percentage nickel for which this can occur?

Using *Figure 3.8* the liquidus at 1140°C gives an alloy with 90% copper and 10% nickel. This is the alloy with the highest percentage nickel which is just liquid at 1140°C.

Eutectic

The equilibrium diagram for the copper–nickel alloy is typical of that given when two components are soluble in each other, both in the liquid and solid states. *Figure 3.9* shows the general form of such an equilibrium diagram.

Figure 3.10 shows the type of equilibrium diagrams produced when the two alloy components are completely soluble in each other in the liquid state but completely insoluble in each other in the solid state. The solid alloy shows a mixture of crystals of the two metals concerned. Each of the two metals in the solid alloy retains its independent identity. At one particular composition, called the *eutectic composition* (marked as E in *Figure 3.10*), the temperature at which

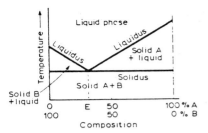

Figure 3.10 Equilibrium diagram for two metals that are completely soluble in each other in the liquid state and completely insoluble in each other in the solid state

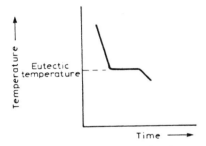

Figure 3.11 Cooling curve for the eutectic composition

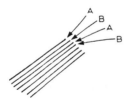

Figure 3.12 The laminar structure of the eutectic

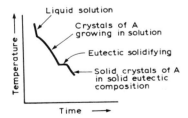

Figure 3.14 Cooling curve for an 80% A – 20% B alloy

solidification starts to occur is a minimum. At this temperature, called the *eutectic temperature*, the liquid changes to the solid state without any change in temperature (*Figure 3.11*). The solidification at the eutectic temperature, for the eutectic composition, has both the metals simultaneously coming out of the liquid. Both metals crystallise together. The resulting structure, known as the *eutectic structure*, is generally a laminar structure with layers of metal A alternating with layers of metal B (*Figure 3.12*).

The properties of the eutectic can be summarised as:

1 Solidification takes place at a single fixed temperature.
2 The solidification takes place at the lowest temperature in that group of alloys.
3 The composition of the eutectic composition is a constant for that group of alloys.
4 It is a mixture, for an alloy made up from just two metals, of the two phases.
5 The solidified eutectic structure is generally a laminar structure.

Figure 3.13 illustrates the sequence of events that occur when the 80% A–20% B liquid alloy is cooled. In the liquid state both metals are completely soluble in each other and the liquid alloy is thus completely homogeneous. When the liquid alloy is cooled to the liquidus temperature, crystals of metal A start to grow. This means that as metal A is being withdrawn from the liquid, the composition of the liquid must change to a lower concentration of A and a higher concentration of B. As the cooling proceeds and the crystals of A continue to grow so the liquid further decreases in concentration of A and increases in concentration of B. This continues until the concentrations in the liquid reach that of the eutectic composition. When this happens solidification of the liquid gives the eutectic structure. The resulting alloy has therefore crystals of A embedded in a structure having the composition and structure of the eutectic. *Figure 3.14* shows the cooling curve for this sequence of events.

Apart from an alloy having the eutectic composition and structure when the alloy is entirely of eutectic composition, all the other alloy compositions in the alloy group show crystals of either metal A or B embedded in eutectic structure material (*Figure 3.15*). Thus for two metals that are completely insoluble in each other in the solid state:

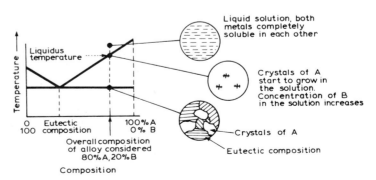

Figure 3.13

48 Alloying of metals

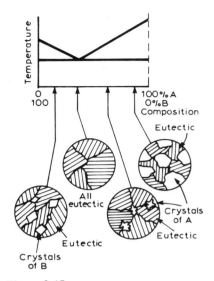

Figure 3.15

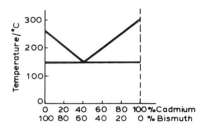

Figure 3.16 Equilibrium diagram for cadmium–bismuth alloys

1 The structure prior to the eutectic composition is of crystals of B in material of eutectic composition and structure.

2 At the eutectic structure the material is entirely eutectic in composition and structure.

3 The structure after the eutectic composition is of crystals of A in material of eutectic composition and structure.

Example
Figure 3.16 shows the equilibrium diagram for alloys of bismuth and cadmium. (a) What is the composition of the eutectic? (b) What will be the structure of a solid 80% cadmium – 20% bismuth alloy? (c) What will be the structure of a solid 20% cadmium – 80% bismuth alloy?

(a) The eutectic composition is 40% cadmium – 60% bismuth.
(b) Crystals of cadmium in eutectic.
(c) Crystals of bismuth in eutectic.

More complex equilibrium diagrams

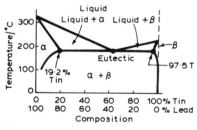

Figure 3.17 Equilibrium diagram for lead–tin alloys

Many metals are neither completely soluble in each other in the solid state nor completely insoluble; each of the metals is soluble in the other to some limited extent. Lead–tin alloys are of this type. *Figure 3.17* shows the equilibrium diagram for lead–tin alloys. The solidus line is that line, starting at 0% tin – 100% lead, between the (liquid + α) and the α areas, between the (liquid + α) and the (α + β) areas, between the (liquid + β) and the (α + β) areas, and between the (liquid + β) and the β areas. The α, the β, and the (α + β) areas all represent solid forms of the alloy. The transition across the line between α and (α + β) is thus a transition from one solid form to another solid form. Such a line is called the *solvus*. *Figure 3.18* shows the early part of *Figure 3.17* and the liquidus, solidus and solvus lines.

Consider an alloy with the composition 15% tin – 85% lead (*Figure 3.18*). When this cools from the liquid state, where both metals are soluble in each other, to a temperature below the liquidus then crystals of the α phase start to grow. The α phase is a solid solution. Solidification becomes complete when the temperature has fallen to that of the solidus. At that point the solid consists entirely of crytals of the α phase. This solid solution consists of 15% tin completely soluble in the 85% lead at the temperature concerned. Further cooling results in no further change in the crystalline structure until the temperature has fallen to that of the solvus. At this temperature the solid solution is saturated with tin. Cooling below this temperature results in tin coming out of solution in

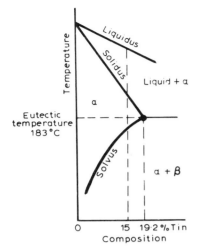

Figure 3.18

another solid solution β. The more the alloy is cooled the greater the amount of tin that comes out of solution, until at room temperature most of the tin has come out of the solid solution. The result is largely solid solution crystals, the α phase, having a low concentration of tin in lead, mixed with small solid solution crystals, the β phase, having a high concentration of tin in lead.

At the eutectic temperature the maximum amount of tin that can be dissolved in lead in the solid state is 19.2% (see *Figure 3.18*). Similarly the maximum amount of lead that can be dissolved in tin, at the eutectic temperature, is 2.5%.

The eutectic composition is 61.9% tin − 38.1% lead. When an alloy with this composition is cooled to the eutectic temperature the behaviour is the same as when cooling to the eutectic occurred for the two metals insoluble in each other in solid state (*Figure 3.13*) except that instead of pure metals separating out to give a laminar mixture of the metal crystals there is a laminar mixture of crystals of the two solid solutions α and β. The α phase has the composition of 19.2% tin − 80.8% lead, the β phase has the composition 97.5% tin − 2.5% lead. Cooling below the eutectic temperature results in the α solid solution giving up tin, due to the decreasing solubility of the tin in the lead, and the β solid solution giving up lead, due to the decreasing solubility of the lead in the tin. The result at room temperature is a structure having a mixture of alpha and beta solid solution, the alpha solid solution having a high concentration of lead and the beta a high concentration of tin.

For alloys having a composition with between 19.2% and 61.9% tin, cooling from the liquid results in crystals of the α phase separating out when the temperature falls below that of the liquidus. When the temperature reaches that of the solidus, solidification is complete and the structure is that of crystals of the α solid solution in eutectic structure material. Further cooling results in α solid solution losing tin. The eutectic mixture has the α part of it losing tin and the β part losing lead. The result at room temperature is a structure having the α solid solution crystals with a high concentration of lead and very little tin and some β precipitate, and the eutectic structure a mixture of α with high lead concentration and β with high tin concentration.

For alloys having a composition with between 61.9% and 97.5% tin, cooling from the liquid results in crystals of the β phase separating out when the temperature falls below that of the liquidus. Otherwise the events are the same as those occurring for compositions between 19.2% and 61.9% tin. The result at room temperature is a structure having the β solid solution crystals with a high concentration of tin and very little lead and some α precipitate, and the eutectic structure a mixture of α with high lead concentration and β with high tin concentration.

For alloys having a composition with more than 97.5% tin present, crystals of the β phase begin to grow when the temperature falls below that of the liquidus. When the temperature falls to that of the solidus solidification becomes complete and the solid consists entirely of β solid solution crystals. Further cooling results in no further change in the structure until the temperature reaches that of

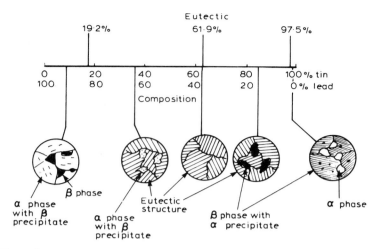

Figure 3.19 Lead–tin alloys

the solvus. At this temperature the solid solution is saturated with lead and cooling below this temperature results in the lead coming out of solution. The result at room temperature, when most of the lead has come out of the solid solution, is β phase crystals having a high concentration of tin mixed with α phase crystals with a high concentration of lead.

Figure 3.19 shows the types of structure that might be expected at room temperature for lead–tin alloys of different compositions.

For many alloys, the phase diagrams are more complex than those already considered in this chapter. The complexity occurs because of the formation of further phases. These can be due to the formation of intermetallic compounds.

Example
Figure 3.20 shows the equilibrium diagram for silver–copper alloys. What will be the structure at room temperature of (a) a 5% copper – 95% silver alloy, (b) a 28.5% copper – 81.5% silver alloy, (c) a 70% copper – 30% silver alloy, (d) a 35% copper – 65% silver alloy?

(a) Solid solution α phase crystals, largely silver, mixed with small solid solution β crystals, largely copper.

(b) This is the eutectic composition and so the entire alloy will be of eutectic structure.

(c) Solid solution β phase crystals, largely copper, mixed with eutectic structure material.

(d) Solid solution β phase crystals, largely copper, mixed with eutectic structure material. The β phase crystals are smaller than in (c).

PROBLEMS

1 What is the difference between a mixture of two substances and a compound of the two?

2 Explain what is meant by the terms soluble, insoluble, liquid solution, solid solution.

Alloying of metals 51

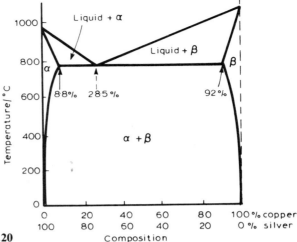

Figure 3.20

3 Sketch the form of cooling curve you would expect when a sample of pure iron is cooled from the liquid to solid state.

4 *Figure 3.21* shows the cooling curves for copper–nickel alloys. Use these to plot the copper–nickel thermal equilibrium diagram.

5 Use either *Figure 3.8* or your answer to Problem 4 to determine the liquidus and solidus temperatures for a 50% copper – 50% nickel alloy.

6 Germanium and silicon are completely soluble in each other in both the liquid and solid states. Plot the thermal equilibrium diagram for germanium–silicon alloys from the following data.

| Alloy | | Liquidus temperature | Solidus temperature |
Germanium %	Silicon %	/°C	/°C
100	0	958	958
80	20	1115	990
60	40	1227	1050
40	60	1315	1126
20	80	1370	1230
0	100	1430	1430

7 Explain what is meant by the liquidus, solidus and solvus lines on a thermal equilibrium diagram.

8 Describe the form of the thermal equilibrium diagrams that would be expected for alloys of two metals that are completely soluble in each other in the liquid state but in the solid state are (a) soluble, (b) completely insoluble, (c) partially soluble in each other.

9 The lead–tin thermal equilibrium diagram is given in *Figure 3.17*

(a) What is the composition of the eutectic?

(b) What is the eutectic temperature?

(c) What will be the expected structure of a solid 40% tin – 60% copper alloy?

(d) What will be the expected structure of a solid 10% tin – 90% copper alloy?

(e) What will be the expected structure of a solid 90% – 10% copper alloy?

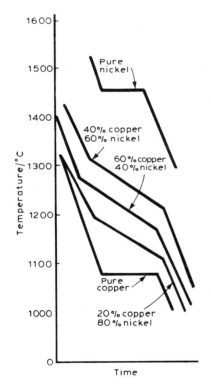

Figure 3.21

4 Ferrous alloys

Objectives: At the end of this chapter you should be able to:

Sketch and identify features on the iron–carbon equilibrium diagram.
Describe the properties of ferrite, cementite, pearlite and austenite.
Describe and explain the structures and properties of hypo- and hyper-eutectoid steels.
Explain the effects of percentage carbon on the properties of carbon steel.
Explain the basic heat treatment processes in terms of the iron–carbon thermal equilibrium diagram.
Describe the processes used for the surface hardening of carbon steel.
Explain the term 'hardenability'.
Explain the term 'limiting ruling section' and recognise the effect of section size on mechanical properties.

THE IRON–CARBON SYSTEM

Pure iron at ordinary temperatures has the body-centred cubic structure. This form is generally referred to as α *iron*. The iron will

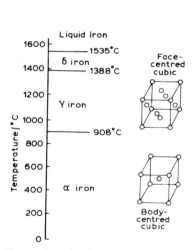

Figure 4.1 The forms of pure iron

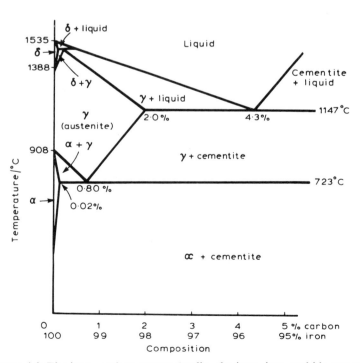

Figure 4.2 The iron–carbon system (really, the iron–iron carbide system)

Ferrous alloys 53

retain this structure up to a temperature of 908°C. At this temperature the structure of the iron changes to face-centred cubic. This is referred to as γ iron. At 1388°C this structure changes again to give a body-centred cubic structure known as δ *iron*. The iron retains this form up to the melting point of 1535°C. *Figure 4.1* summarizes these various changes.

The name *ferrite* is given to the two body-centred cubic forms of iron, i.e. the α and δ forms. The name *austenite* is given to the face-centred cubic form, i.e. the γ form.

Figure 4.2 shows the iron–carbon system thermal equilibrium diagram. The α iron will accept up to about 0.02% of carbon in solid solution. The γ iron will accept up to 2.0% of carbon in solid solution. With these amounts of carbon in solution the α iron still retains its body-centred cubic structure, and the name ferrite, and the γ iron its face-centred structure, and the name austenite (*Figure 4.3*). The solubilities of carbon in iron, both in the austenite and ferrite forms, varies with temperature. With slow cooling, carbon in excess of that which the α or γ solid solutions can hold at a particular temperature, will precipitate. The precipitate is not however as carbon but as iron carbide (Fe_3C), a compound formed between the iron and the carbon. This iron carbide is known as *cementite*. Cementite is hard and brittle.

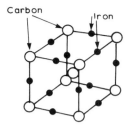

Ferrite, the body centred cubic form, can only accept up to 0.02% carbon in solid solution

(a)

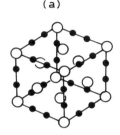

Austenite, the face centred cubic form, can accept up to 2.0% carbon in solid solution

(b)

Figure 4.3 Ferrite and Austenite

Eutectoid Consider the cooling from the liquid of an alloy with 0.80% carbon (*Figure 4.4*). For temperatures above 723°C the solid formed is γ iron, i.e. austenite. This is a solid solution of carbon in iron. At 723°C there is a sudden change to give a laminated structure of

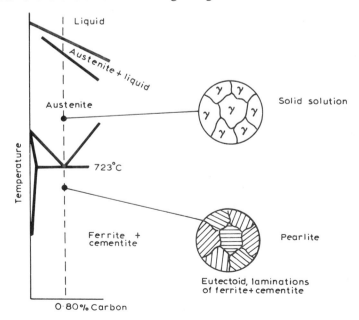

Figure 4.4 Slow cooling of a 0.80% carbon steel

Figure 4.5 Lamellar pearlite × 600 magnification (from Monks, H. A. and Rochester, D. C., *Technician Structure* and *Properties of Metals*, Cassell)

ferrite plus cementite. This structure is called *pearlite* (*Figure 4.5*). This change at 723°C is rather like the change that occurs at a eutectic, but there the change is from a liquid to a solid; here the change is from one solid structure to another. This type of change is said to give a *eutectoid*. The eutectoid structure has the composition of 0.8% carbon – 99.2% iron in this case.

Steels containing less than 0.80% carbon are called hypo-eutectoid steels, those with between 0.80% and 2.0% carbon are called hyper-eutectoid steels.

Figure 4.6 shows the cooling of a 0.4% carbon steel, a hypo-eutectoid steel, from the austenite phase to room temperature. When the alloy is cooled below temperature T_1, crystals of ferrite start to grow in the austenite. The ferrite tends to grow at the grain boundaries of the austenite crystals. At 723°C the remaining austenite changes to the eutectoid structure, i.e. pearlite. The result can be a network of ferrite along the grain boundaries surrounding areas of pearlite (*Figure 4.7*).

Figure 4.8 shows the cooling of a 1.2% carbon steel, a *hyper*-eutectoid steel, from the austenite phase to room temperature. When the alloy is cooled below the temperature T_1, cementite starts to grow in the austenite at the grain boundaries of the austenite crystals. At 723°C the remaining austenite changes to the eutectoid structure, i.e. pearlite. The result is a network of cementite along the grain boundaries surrounding areas of pearlite.

Thus *hypo*-eutectoid carbon steels consist of a *ferrite* network enclosing pearlite and *hyper*-eutectoid carbon steels consist of a *cementite* network enclosing pearlite.

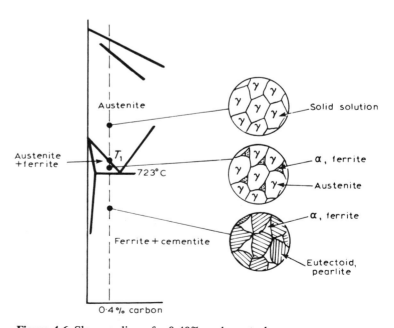

Figure 4.6 Slow cooling of a 0.40% carbon steel

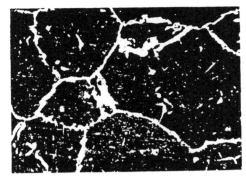

Figure 4.7 A 0.5% carbon steel, slow cooled: shows network of ferrite

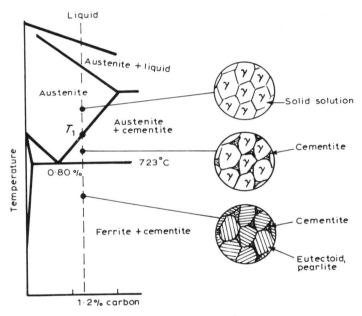

Figure 4.8 Slow cooling of a 1.2% carbon steel

Example
What would be the expected structure of a 1.6% carbon steel if it was slowly cooled from the austenitic state?

A cementite network enclosing pearlite grains.

The effect of carbon content on mechanical properties of steels

Ferrite is a comparatively soft and ductile material. Pearlite is a harder and much less ductile material. Thus the relative amounts of these two substances in a carbon steel will have a significant effect on the properties of that steel. *Figure 4.9* shows how the percentages of ferrite and pearlite change with percentage carbon and also how the mechanical properties are related to these changes. The data refers only to steels cooled slowly from the austenitic state.

Up to the eutectoid composition carbon steel, i.e. for hypo-eutectoid steels, the decreasing percentage of ferrite and the increasing percentage of pearlite results in an increase in tensile strength

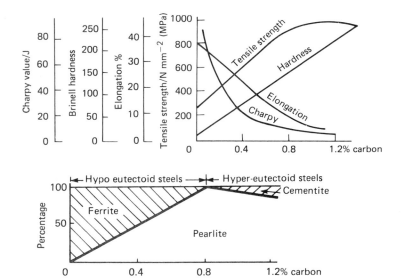

Figure 4.9 The effect of carbon content on the structure and properties of steels

and hardness. The ductility decreases, the elongation at fracture being a measure of this. For hyper-eutectoid steels, increasing the amount of carbon decreases the percentage of pearlite and increases the percentage of cementite. This increases the hardness but has little effect on the tensile strength. The ductility also changes little.

The limitations to the use of plain carbon steels are that high strength steel, up to about 600 N mm^{-2} (MPa) can be obtained only by increasing the carbon content to such a level that the material becomes brittle. Steels with more than 0.3% carbon can be hardened by rapid cooling, i.e. quenching, from austenite to give martensite instead of ferrite.

ALLOY STEELS

The term *alloy steel* is used to describe those steels to which one or more alloying elements, in addition to carbon, have been added deliberately in order to modify the properties of the steel. In general the steels have less than 1% carbon and the additions are, in the case of manganese, more than 1.5% or, in the case of silicon, more than 0.5%.

There are a number of ways in which alloying elements can have an effect on the properties of the steel:

1 They may go into solid solution, giving an increase in strength. In *low-alloy steels*, i.e. those with less than about 3 to 4% of alloying elements other than carbon, the alloying elements enter into solid solution with the ferrite and so give a stronger steel which still has good ductility. Silicon is an example of such an alloying element.

2 They may form stable, hard, carbides or nitrides which may, if in an appropriate form such as fine particles, increase the strength and hardness. Manganese, chromium and tungsten have this effect.

3 They could cause the breakdown of cementite and lead to the

presence of graphite in the structure. Silicon and nickel have this effect. The result is a decrease in strength and hardness. For this reason silicon and nickel are not used with high carbon steels.

4 They may lower the temperature at which austenite is formed on heating the steel. Manganese, nickel, copper and cobalt have this effect. The lowering of this temperature means a reduction in the temperature to which the steel has to be heated for hardening by quenching. If a sufficiently high percentage of one of these elements is added to the steel the transformation temperature to austenite may be decreased to such an extent that the austenite is retained at room temperature. With manganese about 11 to 14% produces what is known as an *austenitic steel*. Such steels have relatively good hardness combined with ductility and so are tough.

5 They may increase the temperature at which austenite is formed on heating the steel. Chromium, molybdenum, tungsten, vanadium, silicon and aluminium have this effect. This raises the temperature to which the steel has to be heated for hardening. If a sufficiently high percentage of one of these elements is added to the steel the transformation from ferrite to austenite may not take place before the steel reaches its melting point temperature. Such steels are known as *ferritic steels*, 12 to 25% of chromium with a 0.1% carbon steel gives such a type of steel. Such a steel cannot be hardened by quenching.

6 They could change the critical cooling rate. The *critical cooling rate* is the minimum rate of cooling that has to be used if all the austenite is to be changed into martensite. With the critical cooling rate, or a higher rate of cooling, the steel has maximum hardness. With slower rates of cooling a less hard structure is produced. Most alloying elements reduce the critical cooling rate. The effect of this is to make air or oil quenching possible, rather than water quenching. It also increases the hardenability of steels.

7 They can influence grain growth. Some elements accelerate grain growth while others decrease grain growth. The faster grain growth leads to large grain structures and consequentially to a degree of brittleness. The slower grain growth leads to smaller grain size and so to an improvement in properties. Chromium accelerates grain growth and thus care is needed in heat treatment of chromium steels to avoid excessive grain growth. Nickel and vanadium decrease grain growth.

8 They can affect the machinability of the steel. Sulphur and lead are elements that are used to improve the chip formation properties of steels.

9 They can improve the corrosion resistance. Some elements promote the production of adherent oxide layers on the surfaces of the steel and so improve its corrosion resistance. Chromium is particularly useful in this respect. If it is present in a steel in excess of 12% the steel is known as *stainless steel* because of its corrosion resistance. Copper is also used to improve corrosion resistance.

The following list indicates the main effects of the commonly used alloying elements. As will have been apparent from the preceding discussion an alloying element generally has more than one way of affecting the properties of the alloy.

Element	Typical amount	Main effects on properties
Aluminium	0.95 to 1.3%	Aids nitriding. Restricts grain growth.
Chromium	0.5 to 2%	Increases hardenability.
	4 to 18%	Improves corrosion resistance.
Copper	0.1 to 0.5%	Improves corrosion resistance.
Lead	Less than 0.2%	Improves machinability.
Manganese	0.2 to 0.4%	Combines with the sulphur in the steel to reduce brittleness.
	1 to 2%	Increases hardenability.
Molybdenum	0.1 to 0.5%	Inhibits grain growth. Improves strength and toughness.
Nickel	0.3 to 65%	Improves strength and toughness. Increases hardenability.
Phosphorus	0.05%	Improves machinability.
Silicon	0.2 to 2%	Increases hardenability. Removes oxygen in steel making.
Sulphur	Less than 0.2%	Improves machinability.
Tungsten		Hardness at high temperatures.
Vanadium	0.1 to 0.3%	Restricts grain growth. Improves strength and toughness.

CRITICAL CHANGE POINTS

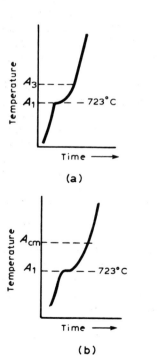

Figure 4.10 Heating curves for (a) a hypo-eutectoid steel, (b) a hyper-eutectoid steel

If water is heated and a graph plotted of temperature with time there will be found to be two discontinuities in the graph where the temperature does not continue to rise at a steady rate despite the heat being supplied at a steady rate. *Figure 3.5* shows one of these discontinuities. They occur at 0°C and 100°C when the state of the water changes. When carbon steels are heated, similar discontinuities occur in the temperature with time graph. The temperatures at which there are changes in the rate of temperature rise, for a constant rate of supply of heat, for steels are known as *arrest points* or *critical points*.

Figure 4.10a shows a heating curve for a hypo-eutectoid steel. The lower critical temperature A_1 is the same for all carbon steels and is the temperature 723°C at which the eutectoid change occurs. For the hypo-eutectoid steel this temperature marks the transformation from a steel with a structure of ferrite and cementite to one of ferrite and austenite. The upper critical temperature A_3 depends on the carbon content of the steel concerned and marks the change from a structure of ferrite and austenite to one solely of austenite. For a 0.4% carbon steel this would be temperature T_1 in *Figure 4.6*.

The heating curve for a hyper-eutectoid steel (*Figure 4.10b*) has the same lower critical temperature A_1 of 723°C, marking the transformation from a steel with a structure of ferrite and cementite to one of austenite and cementite. The upper critical temperature A_{cm} marks the transformation from a structure of austenite and cementite to one solely austenite. For a 1.2% carbon steel this would be temperature T_1 in *Figure 4.8*.

Cooling curves give critical points differing slightly from those produced by heating, cooling giving lower values than heating. The heating critical points are generally denoted by the inclusion of the letter c, the abbreviation for the French word for heating — *chauffage*, i.e. Ac_1, Ac_3, Ac_{cm}. The cooling critical points are denoted by the inclusion of the letter r (the abbreviation for the

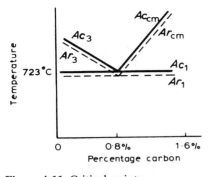

Figure 4.11 Critical points

French word for cooling – *refroidissement*), i.e. Ar_1, Ar_3, Ar_{cm}. The letter 'A' used with the critical points stands for the term 'arrest'.

Figure 4.11 shows a graph of the critical point temperatures against the percentage carbon in the steel. The graph is restricted to those carbon percentages that result in steels. The graph is, in fact, the iron–carbon diagram given in *Figure 4.2*. The critical point graph is the thermal equilibrium graph if the heating and cooling graphs were obtained as the result of very slow heating and cooling rates.

HEAT TREATMENT OF STEEL

Heat treatment can be defined as the controlled heating and cooling of metals in the solid state for the purpose of altering their properties according to requirements. Heat treatment can be applied to steels to alter their properties by changing grain size and the form of the constituents present.

Annealing is the heat treatment used to make a steel softer, and more ductile, remove stresses in the material and reduce the grain size. One form of the annealing process is called *full annealing*. In the case of hypo-eutectoid steels, this involves heating the material to a temperature above the A_3 temperature, holding at that temperature for a period of time and then very slowly cooling it. Typically, the material is heated to about 40°C above the A_3 temperature which has the effect of converting the structure of the steel to austenite. Slow cooling leads to the conversion of the austenite to ferrite and pearlite. The result is a steel in as soft a condition as possible.

A different process has to be used for hyper-eutectoid steels in that heating them to above the A_{cm} temperature turns the entire steel structure into austenite but slow cooling from that temperature results in the formation of a network of cementite surrounding the pearlite. The cementite is brittle and has the effect of making the steel relatively brittle. To make the steel soft the original heating is to a temperature only about 40°C above the A_1 temperature. This converts the structure into austenite plus cementite. Slow cooling from this temperature gives as soft a material as is possible, but not however as soft as the full annealing process gives when applied to hyper-eutectoid steels.

Sub-critical annealing, sometimes referred to as *process annealing*, is often used during cold working processes where the material has to be made more ductile for the process to continue. The process involves heating the material to a temperature just below the A_1 temperature, holding it at that temperature for a period of time and then cooling it at a controlled rate, generally just in air rather than cooling in the furnace as with full annealing. This process leads to no change in structure, no austenite being produced, but just a re-crystallisation. The process is used for steels having up to about 0.3% carbon. When sub-critical annealing is applied to steels having higher percentages of carbon the effect of the heating is to cause the cementite to assume spherical shapes, hence the process is often

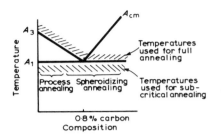

Figure 4.12 Annealing temperatures

referred to as *spheroidising annealing*. This spheroidising results in a greater ductility and an improvement in machinability. *Figure 4.12* summarises the range of annealing processes.

Normalising is a heat treatment process similar to full annealing, and is applied to hypo-eutectoid steels. The steel is heated to about 40°C above the A_3 temperature, held at this temperature for a short while, and then cooled freely in air. The effect of the heating is to form an austenite structure. The cooling rate is however much faster than that with the annealing process. The result is a finer grain structure, ferrite and pearlite. This finer grain size improves the machinability and gives a slightly harder and greater strength material than that given with full annealing.

Both annealing and normalising involve relatively slow cooling of the steel, but what happens if a steel is cooled very quickly, i.e. quenched? With very slow cooling there is time for diffusion to occur in the solid solutions, with quenching there is not time for such events. If a hypo-eutectoid steel is heated to 40°C above the A_3 temperature all the structure becomes austenitic. Very rapid cooling of this structure does not allow sufficient time for the ferrite structure to be produced, i.e. there is not enough time for the austenite to give up its surplus carbon and so produce ferrite. The result is a new structure called *martensite*. Martensite is a very hard structure, hence the result of such a process is a much harder material. The *hardening* process is thus a sequence of heating to produce an austenitic state and then rapid cooling to produce martensite (*Figure 4.13*).

If a steel is cooled at, or faster than, a certain minimum rate, called the *critical cooling rate*, all the austenite is changed into martensite. This gives the maximum hardness. If the cooling is slower than this critical cooling rate a less hard structure is produced. The critical cooling rate depends on the percentage of carbon in the carbon steel, being lower the greater this percentage. The high cooling rates required for low carbon steels mean that steels with less than about 0.3% carbon cannot effectively be hardened.

The rate of cooling depends on the quenching medium used. Water is a commonly used quenching medium and gives a high cooling rate. However, distortion and cracking may be caused by this high cooling rate. Oil gives a slower cooling rate, but brine gives a rate even higher than water. The brine may however give rise to corrosion problems with the steel. Sodium or potassium hydroxide solution is sometimes used for very high cooling rates. In order to minimise the distortion that can be produced during the quenching process, long items should be quenched vertically, flat sections edgeways. To prevent bubbles of steam adhering to the steel during the quenching, and giving rise to different cooling rates for part of the object, the quenching bath should be agitated. Thick objects offer special problems in that the outer parts of the object cool more rapidly than the inner parts. The result can be an outer layer of martensite and an inner core of pearlite, which gives a variation in mechanical properties between the inner and outer parts. This effect is known as the *mass effect*.

Hypo-eutectoid steels are generally hardened by quenching from about 40°C above the A_3 temperature, though steels with less than

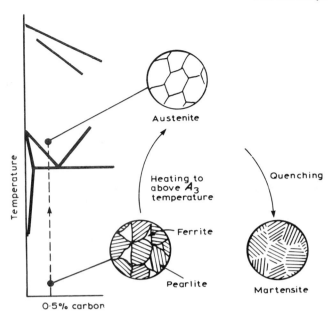

Figure 4.13 Hardening

about 0.3% carbon cannot be hardened very effectively. Hypereutectoid steels are hardened by quenching from about 40°C above the A_1 temperature. This is because to heat the steel above the A_{cm} temperature and then quench it gives rise to a network of cementite and so a brittle structure.

Tempering is the name given to the process in which a steel, hardened as a result of quenching, is reheated to a temperature below the A_1 temperature in order to modify the structure of the steel. The result is an increase in ductility at the expense of hardness and strength. The degree of change obtained depends on the temperature to which the steel is reheated, the higher the tempering temperature the lower the hardness but the greater the ductility (*Figure 4.14*). By combining hardening with tempering an appropriate balance of mechanical properties can be achieved for a steel.

The following tempering temperatures are used to obtain the required properties for the components concerned.

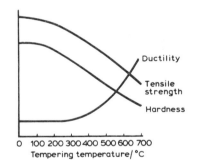

Figure 4.14 The effect of tempering temperature on the properties of a hardened steel

Tempering temperature/°C	Component
200	Scribers
220	Hacksaw blades
230	Planing and slotting tools
240	Drills, milling cutters
250	Taps, shear blades, dies
260	Punches, reamers
270	Axes, press tools
280	Cold chisels, wood chisels
290	Screw drivers
300	Saws, springs

The lower the temperature at which tempering occurs, the harder the product. Thus scribers need to be hard and have high abrasion resistance. Springs, however, do not need to be so hard but more 'springy'. The scribers may be relatively brittle, the springs will be tougher and much less brittle.

Internal stresses in a material can be relieved to some extent by heating it to a temperature say 50° to 100°C below the A_1 temperature. The material is then usually air cooled from that temperature. *Stress relief* by this method is used with welded components before machining, parts requiring machining to accurate dimensions, castings before machining, etc.

Example
The lower critical point A_1 of a 0.4% carbon steel is 723°C, the upper critical point A_3 being 810°C. To what temperature should the steel be heated for full annealing?

A 0.4% carbon steel is a hypo-eutectoid steel and thus it should be heated to about 40°C above the A_3 temperature, i.e. about 855°C.

Example
The lower critical point A_1 of a 0.5% carbon steel is 723°C, the upper critical point A_3 being 765°C. To what temperature should the steel be heated for normalising?

The required temperature is about 40°C above the A_3 temperature, i.e. about 805°C.

Example
What tempering temperature should be used for a hammer if it is to be hard?

A temperature of about 230°C is generally used. A hard component requires a low tempering temperature.

Heat treatment cycles

A heat treatment cycle consists normally of three parts:

1 Heating to get the object to the required temperature for the changes in structure within the material to occur.

2 Holding at that temperature for a long enough time for the entire material to reach the required temperature and the structural changes to occur throughout the entire material. Grain size is affected by the length of holding time.

3 Cooling, with the rate of cooling being controlled since it affects the structure and properties of the material.

Figure 4.15 shows the types of heat treatment cycles used for annealing, normalising, hardening and tempering hypo- and hyper-eutectoid steels.

The effects of heat treatment

The following table shows the basic mechanical properties of annealed carbon steels.

Composition	Condition	Tensile strength /N mm^{-2} (MPa)	Elongation /%	Hardness HB
0.2% carbon	Annealed	430	37	115
0.4% carbon	Annealed	510	30	145
0.6% carbon	Annealed	660	23	190
0.8% carbon	Annealed	800	15	220
1.0% carbon	Annealed	840	12	240

Ferrous alloys 63

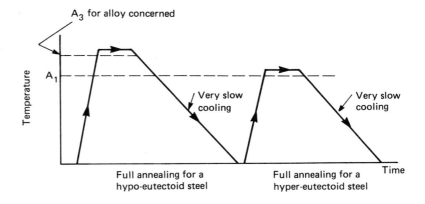

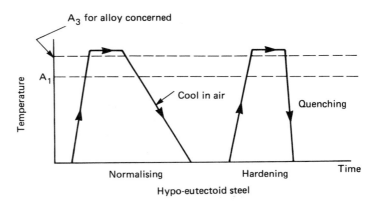

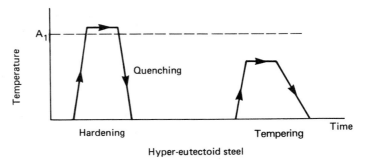

Figure 4.15 Heat treatment cycles

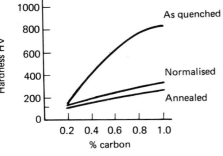

Figure 4.16 Hardness of carbon steels

A normalised carbon steel has a tensile strength approximately 20% higher than the same steel when annealed, with less ductility and greater hardness. Thus, for example, 0.4% carbon steel has, when normalised, a tensile strength of 580 N/mm² (MPa), a percentage elongation of 25% and a hardness of 165 HB.

Steels with less than 0.3% carbon cannot be hardened effectively. With steels with higher carbon contents the effect of quenching is to increase the hardness and tensile strength but decrease the percentage elongation. *Figure 4.16* shows the effect on the hardness of quenching. The effect is greatest for the higher percentage carbon steels. The effect of tempering on the properties of a quenched carbon steel will depend on the quenching temperature, see *Figure 4.14*. Typically, quenching and then tempering a 0.40% carbon steel will increase the tensile strength by about 40 to 50% to about 750 N/mm² (MPa), increase the hardness by about the same percentage to round about 200 to 220 HB, and decrease the percentage elongation to about 16%.

HARDENABILITY

When a block of steel is quenched the surface can show a different rate of cooling to that of the inner core of the block. Thus the formation of martensite may differ at the surface from the inner core. This means a difference in hardness. *Figure 4.17* shows how the hardness varies with depths for a number of different diameter bars when quenched in water and in oil.

With the water quenching, the hardness in the inner core is quite significantly different from the surface hardness. The larger diameter bars also show lower surface and core hardness as their increased mass has resulted in a lower overall rate of cooling. With the oil quenching the cooling rates are lower than those of the water-cooled bars and thus the hardness values are lower. The hardness value for a fully martensitic 0.48% plain carbon steel is about 60 HRC. Thus, as the values in *Figure 4.17* indicate, the quenched bars are not entirely martensitic.

The term *hardenability* is used as a measure of the depth of hardening introduced into a steel section by quenching (not to be

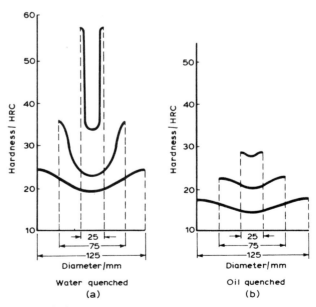

Figure 4.17 Variation of hardness with depth in quenched bars of different diameters for a 0.48% plain carbon steel

confused with hardness). Hardenability is measured by the response of a steel to a standard test.

The hardenability of steel depends on:
1 The quenching medium and the method of quenching.
2 The size of section of the steel item.
3 The composition of the steel.

The quenching medium, method of quenching, size and section of the steel all affect the rate of cooling. The term *mass effect* is used to describe the effect of size of section on the hardenability. The term *ruling section* is used to specify the limit of thickness of a section which can be hardened if the specified mechanical properties are to be obtained. If the ruling section is exceeded then the rate of cooling in the centre of the section may be insufficient to convert it to martensite.

Certain alloying elements added to the steel can change the rate of cooling needed for martensite to be produced. Thus chromium, nickel and manganese reduce the rate of cooling needed and mean that sections of larger size can be fully hardened.

Surface hardening of carbon steels

There is often the need for the surface of a piece of steel to be hard, i.e. wear resistant, without the entire component being made hard and so often too brittle. Several methods are available for surface hardening. For carbon steels there are two methods:
1 Selective heating of the surface layers;
2 Changing the carbon content of the surface layers.

One method of selective heating is called *flame hardening*, which involves heating the surface of a steel with an oxy-acetylene flame and then quenching it before the inner parts of the component have reached the surface temperature (*Figure 4.18*). Another method places the steel component within a coil which carries a high-frequency current, producing currents in the surface layers of the steel by electromagnetic induction (*Figure 4.19*). These currents can heat the surface layers so quickly that the inner parts of the steel component do not reach a high enough temperature for any hardening of them to occur when the component is quenched. This method is called *induction hardening*.

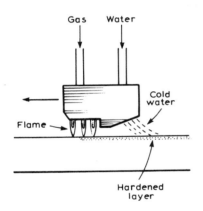

Figure 4.18 A burner with cooling water for flame hardening

Both these selective heating methods require the surface layers of the steel to be brought to above the A_3 temperature while the inner parts of the component remain at a temperature significantly lower. The A_3 temperature results in the surface layers becoming austenitic. When the material is quenched the surface layers become martensite, a much harder material than the inner parts of the component for which the change has not occurred. Steels for surface hardening by these methods require more than about 0.4% carbon content. The hardness produced is typically about 600 HB.

Low-carbon steels can be surface hardened by increasing the carbon content of the surface layers. This method is generally known as *case hardening*. There are a number of stages in this process. In the first stage, *carburising*, the carbon is introduced into the surface layers. With *pack carburising* the steel component is heated to above the A_3 temperature, say 900°C for a 0.1% carbon alloy, while packed in charcoal and barium carbonate. The length of

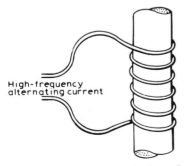

Figure 4.19 Induction hardening of a tube or rod

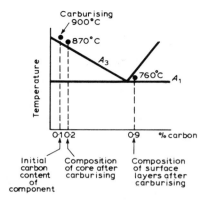

Figure 4.20 Case hardening

time the component is kept at this temperature determines the depth to which the carbon penetrates the surface, this being referred to as the depth of case. In *gas carburising* the component is heated to above the A_3 temperature in a furnace in an atmosphere of a carbon-rich gas. With *cyanide carburising* the component is heated in a bath of liquid sodium cyanide. The result by all these methods is an increase in carbon in the surface layers of the steel component. Thus while the surface layers may have, say, 0.9% carbon the inner core might be 0.2%, some carbon often reaching the core.

The lengthy heating during the carburising treatment causes grain growth, both in the core and the surface layers. Because of the different carbon contents of the core and the surface layers, two heat treatment processes are needed to refine the grains in the core and the surface layers. The core is first refined. The component is heated to above the A_3 temperature for the carbon content of the core (*Figure 4.20*). For a core with 0.2% carbon this is a temperature of about 870°C. The component is then quenched in oil. This treatment results in a fine grain core. The surface layers after this treatment are, however, rather coarse, brittle martensite, because the 870°C temperature is well above the critical temperature for a 0.9% carbon steel. A second heat treatment is then given, specifically to refine the surface layers. The component is heated to above the A_1 temperature, i.e. about 760°C, and then water quenched. The result is fine martensite, which is hard in the surface, while the core is tempered by the treatment. A low temperature tempering is then usually employed about 150°C, to relieve internal stresses produced by the treatment.

The case hardening treatment might thus be:

1 Carburise by heating to 950°C in a carbon-rich environment,
2 Core refine by heating to 870°C and quench in oil,
3 Refine the surface layers, and so case harden, by heating to 750°C and quench in water,
4 Temper at 150°C to relieve internal stresses.

The hardness produced by such a method is typically of the order of 850 HB.

Nitriding involves changing the surface composition of a steel by diffusing nitrogen into it; hard compounds, nitrides, are produced. The process is used with those alloy steels that contain elements that form stable nitrides, e.g. aluminium, chromium, molybdenum, tungsten, vanadium. Prior to the nitriding treatment the steel is hardened and tempered to the properties required of the core. The tempering temperature does however need to be in the region 560°C to 750°C. The reason for this is that the nitriding process requires a temperature up to about 530°C, and this must not be greater than the tempering temperature, as the nitriding process would temper the steel and so change the properties of the core.

Unlike carburising, nitriding is carried out at temperatures below the stable austenitic state. The process consists of heating a component in an atmosphere of ammonia gas and hydrogen, the temperatures being of the order of 500°C to 530°C. The time taken for the nitrogen to react with the elements in the surface of the steel is often as much as 100 hours. The depth to which the nitrides are formed in the steel depends on the temperature and the time allowed for the reaction. Even with such long times the depth of hardening is unlikely to exceed about 0.7 mm. After the treatment the component is allowed to cool slowly in the ammonia/hydrogen atmosphere. With most nitriding conditions a thin white layer of iron nitrides is formed on the surface of the component. This layer adversely affects the mechanical properties of the steel, being brittle and generally containing cracks. It is therefore removed by mechanical means or chemical solutions.

Because with nitriding no quenching treatments are involved, cracking and distortion are less likely than with other surface hardening treatments. Very high surface hardnesses can be obtained with special alloys. The hardness is retained at temperatures up to about 500°C, whereas that produced by carburising tends to decrease and the surface becomes softer at temperatures of the order of 200°C. The capital cost of the plant is however higher than that associated with pack carburising.

Carbonitriding is the name given to the surface hardness process in which both carbon and nitrogen are allowed to diffuse into a steel when it is in the austenitic–ferritic condition. The component is heated in an atmosphere containing carbon and ammonia and the temperatures used are about 800°C to 850°C. The nitrogen inhibits the diffusion of carbon into the steel and, with the temperatures and times used being smaller than with carburising, this leads to relatively shallow hardening. Though this process can be used with any steel that is suitable for carburising it is used generally only for mild steels or low alloy steels.

Ferritic nitrocarburising involves the diffusion of both carbon and nitrogen into the surface of a steel. The treatment involves a temperature below the A_1 temperature when the steel is in a ferritic condition. A very thin layer of a compound of iron, nitrogen and carbon is produced at the surfaces of the steel. This gives excellent wear and anti-scuffing properties. The process is mainly used on mild steel in the rolled or normalised condition.

The following table compares the main surface hardening treatments. In comparing the processes the term *case depth* is used, which is defined graphically (*Figure 4.21*) in terms of a graph of carbon or nitrogen content against depth under the surface of the

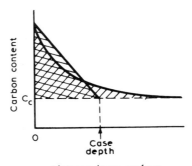

Figure 4.21 Defining case depth

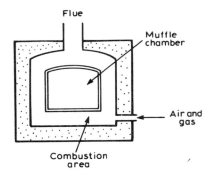

Figure 4.22 The basic form of a gas-fired muffle furnace

steel. A straight line is drawn on the graph such that it passes through the surface carbon content value and so that the area between the line and the core carbon content C_c is the same as that between the actual carbon content graph curve and the core carbon content C_c. Where this line meets the core carbon content line gives the case depth.

Process	Temperature /°C	Case depth /mm	Case hardness /HRC	Main user
Pack carburising	810–1100	0.25–3	45–65	Low carbon and carburising alloy steels. Large case depths, large components.
Gas carburising	810–980	0.07–3	45–65	Low carbon and carburising alloy steels. Large numbers of components.
Cyaniding	760–870	0.02–0.7	50–60	Low carbon and light alloy steels. Thin case.
Nitriding	500–530	0.07–0.7	50–70	Alloy steels. Lowest distortion.
Carbo-nitriding	700–900	0.02–0.7	50–60	Low carbon and low alloy steels.
Flame hardening	850–1000	Up to 0.8	55–65	0.4 to 0.7% carbon steels, selective heating.
Induction hardening	850–1000	0.5–5	55–65	0.4 to 0.7% carbon steels, selective heating.

Example

A steel with 0.6% carbon is to be surface hardened. Which type of process, selective heating or carburising, would be suitable?

Carburising is used with steels having less than about 0.2% carbon while selective heating requires steel with 0.4% or more carbon. Thus selective heating is the suitable method.

HEAT TREATMENT EQUIPMENT

The industrial heat treatment of metals requires:

1 Furnaces to heat the components to the required measured temperature;

2 Quenching equipment.

An important consideration before the choice of furnace is made is whether the atmosphere in which the components are heated is to be oxidising, i.e. generally air, or some inert gas to prevent oxidisation. A *muffle furnace* (*Figure 4.22*) is one in which the components are contained in a chamber, the so-called muffle chamber, and the heat is supplied externally to the chamber. This enables the atmosphere in the chamber to be controlled. A non-muffle furnace has no control over the atmosphere surrounding the components.

Heating an iron alloy in an oxidising atmosphere can result in the carbon in the outer layers of the material combining with the oxygen to give carbon dioxide, which escapes with the flue gases and so leaves the outer layers of the alloy with less carbon than the inner layers. This effect is known as *decarburisation*. Scaling and tarnishing can also occur when an alloy is heated in an oxidising atmosphere. Heating in a non-oxidising atmosphere can eliminate costly cleaning processes.

The *salt-bath furnace* consists essentially of a pot containing a molten salt, which may be heated directly by an electric current being passed through it or by external heating of the pot. Salt-bath furnaces give an even and rapid heating of components immersed in the salt. By suitable choice of the salt, carburisation can be promoted and so surface hardening occurs. No oxidisation takes place for the components immersed in the salt as air is excluded.

Quenching may involve the immersion of the hot components in water, brine, oil or sodium or potassium hydroxide solution. The quenching tanks are designed so that the quenching liquid remains, as nearly as possible, at a constant temperature. In the case of brine or oil baths, these may stand in large tanks through which water flows, the water carrying the heat away and so keeping the quenching liquids at a near constant temperature during the quenching.

There are a number of hazards during heat treatment processes:

1 Hot metals are being handled, so care should be taken in approaching any piece of metal during the heat treatment process.

2 Quenching baths and gas fired furnaces give off dangerous fumes. The heat treatment area needs to be well ventilated and fume extraction systems should be used.

3 Fire hazards are present with furnaces and also oil quenching tanks. The immersion of a piece of hot metal in oil leads to some of the oil being vaporised and the possibility of a fire. If this happens the air supply to the tank should be cut off by covering the tank with a lid.

4 Explosions are possible with gas-fired furnaces if the furnace becomes charged with a gas–air mixture before the lighter is operated. The lighter should be operated before the gas is introduced into the combustion chamber.

5 Some of the materials used in heat treatment are poisonous, e.g. sodium cyanide is used for salt-bath furnaces.

PROBLEMS

1 Sketch and label the steel section of the iron–carbon system, using the terms austenite, ferrite and cementite.

2 How do the structures of *hypo*-eutectoid and *hyper*-eutectoid steels differ at room temperature as a result of their being slowly cooled from the austenitic state?

3 Describe the form of the microstructure of a slowly cooled steel having the eutectoid structure.

4 What would be the expected structure of a 1.1% carbon steel if it were cooled slowly from the austenitic state?

5 Explain the terms critical points A_1, A_3 and A_{cm}.

6 Explain how the percentage of carbon present in a carbon steel affects the mechanical properties of the steel.

7 Describe the following heat treatments applied to steels.
   (a) Full annealing.
   (b) Normalising.
   (c) Hardening.
   (d) Tempering.

8 What will be the form of the microstructure of a 0.5% carbon steel after the following treatment: Heat and soak at 805°C and then a very slow cool to room temperature?

9 How would the answer to Problem 8 have differed if the steel, instead of being slowly cooled from 805°C to room temperature, had been quenched?

10 Explain how a 0.4% carbon steel would be hardened. Give details of the temperatures involved.

11 Describe the difference between full annealing and normalising.

12 State what is meant by the critical cooling rate.

13 How does increasing the temperature at which a carbon steel is tempered change the final properties of the steel?

14 How would the mechanical properties of 0.6% carbon steels differ if the following heat treatments were applied?
  (a) Heat and soak at 800°C and then quench in cold water.
  (b) Heat and soak at 800°C and then quench in oil.
  (c) Heat and soak at 800°C and then slowly cool in the furnace.

15 A carbon steel with 1.1% carbon is to be given a full annealing treatment. What temperature and cooling rate are necessary for such a treatment?

16 Why are cylindrical objects quenched vertically?

17 A cold chisel is tempered at a temperature of 280°C while a scriber is tempered at 200°C. How does the hardness of the steel differ for the two items as a result of the differing tempering temperatures? Why are the components required to have different hardnesses? What would happen if there was an error and the tempering temperature for the cold chisel was as high as 380°C?

18 Explain, by reference to an iron–carbon thermal equilibrium diagram, the procedures used for the case hardening of a low carbon steel.

19 Describe the functions of the muffle and salt-bath furnaces?

20 State the form of heat treatments needed to effect the following changes:
  (a) a 0.2% carbon steel to be made as soft as possible;
  (b) a 1.0% carbon steel to be made as soft as possible;
  (c) a 0.4% carbon steel to be made as hard as possible;
  (d) a 0.2% carbon steel to be case hardened.

21 Explain why steels below about 0.3% carbon cannot be significantly hardened by quenching unless other alloying elements are added.

22 Explain the terms 'hardenability', 'mass effect' and 'ruling section'.

23 In materials specifications the ruling section is quoted. Why?

24 Explain what is meant by a heat treatment cycle and the functions of the heating, holding and cooling parts of the cycle.

25 Explain how flame hardening leads to the surface hardening of a steel.

26 Explain how the hardening of a steel surface can be affected by changing the surface composition.

27 Compare the surface hardening processes of nitriding and carburising.

5 Non-ferrous alloys

Objectives: At the end of this chapter you should be able to:

Describe the properties of aluminium.
Describe the properties and uses of common aluminium alloys.
Describe the effect of composition and temperature on the equilibrium structure of aluminium alloys containing up to 8% copper.
Explain the precipitation hardening of an aluminium–copper alloy.
Describe the properties of copper.
Describe the properties and uses of common copper alloys.
Describe the basic characteristics of magnesium, nickel, titanium and zinc alloys.

ALUMINIUM

Pure aluminium has a density of 2.7×10^3 kg m^{-3}, compared with that of 7.9×10^3 kg m^{-3} for iron. Thus for the same size component the aluminium version will be about one third of the mass of an iron version. Pure aluminium is a weak, very ductile, material. It has an electrical conductivity about two thirds that of copper but weight for weight is a better conductor. It has a high thermal conductivity. Aluminium has a great affinity for oxygen and any fresh metal in air rapidly oxidises to give a thin layer of the oxide on the metal surface. This surface layer is not penetrated by oxygen and so protects the metal from further attack. The good corrosion resistance of aluminium is due to this thin oxide layer on its surface.

High purity aluminium is too weak a material to be used in any other capacity than as a lining for vessels. It is used in this way to give a high corrosion resistant surface. High purity aluminium is 99.5%, or more, aluminium.

Commercial purity aluminium, 99.0 to 99.5% aluminium, is widely used as aluminium foil for sealing milk bottles and for thermal insulation, as kitchen foil for cooking. The presence of a relatively small percentage of impurities in aluminium considerably increases the tensile strength and hardness of the material.

The mechanical properties of aluminium depend not only on the purity of the aluminium but also upon the amount of work to which it has been subject. The effect of working the material is to fragment the grains. This results in an increase in tensile strength and hardness and a decrease in ductility. By controlling the amount of working, different degrees of strength and hardness can be produced. These are said to be different *tempers*. The properties of aluminium may thus, for example, be referred to as that for the annealed condition, the half-hard temper and the fully hardened temper.

The table shows the types of properties that might be obtained with aluminium.

Composition %	Condition	Tensile strength /N mm^{-2} (MPa)	Elongation %	Hardness /HB
99.99	Annealed	45	60	15
	Half hard	82	24	22
	Full hard	105	12	30
99.8	Annealed	66	50	19
	Half hard	99	17	31
	Full hard	134	11	38
99.5	Annealed	78	47	21
	Half hard	110	13	33
	Full hard	140	10	40
99.0	Annealed	87	43	22
	Half hard	120	12	35
	Full hard	150	10	42

ALUMINIUM ALLOYS

Aluminium alloys can be divided into two groups:
1 Wrought alloys,
2 Cast alloys.

Each of these groups can be divided into two further groups:
(a) Those alloys which are not heat treatable,
(b) Those alloys which are heat treated.

The term *wrought material* is used for a material that is suitable for shaping by working processes, e.g. forging, extrusion, rolling. The term *cast material* is used for a material that is suitable for shaping by a casting process.

The non-heat treatable wrought alloys of aluminium do not significantly respond to heat treatment but have their properties controlled by the extent of the working to which they are subject. A range of tempers is thus produced. Common alloys in this category are aluminium with manganese or magnesium. A common aluminium–manganese alloy has 1.25% manganese, the effect of this manganese being to increase the tensile strength of the aluminium. The alloy still has a high ductility and good corrosion properties. This leads to uses such as kitchen utensils, tubing and corrugated sheet for building. Aluminium–magnesium alloys have up to 7% magnesium. The greater the percentage of magnesium the greater the tensile strength (*Figure 5.1*). The alloy still has good ductility. It has excellent corrosion resistance and thus finds considerable use in marine environments, e.g. constructional materials for boats and ships.

The heat treatable wrought alloys can have their properties changed by heat treatment. Copper, magnesium, zinc and silicon are common additions to aluminium to give such alloys. *Figure 5.2* shows the thermal equilibrium diagram for aluminium–copper alloys. When such an alloy, say 3% copper – 97% aluminium, is slowly cooled the structure at about 540°C is a solid solution of the α

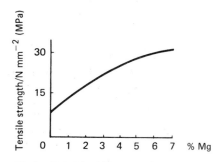

Figure 5.1 The effect on the tensile strength of magnesium content in annealed aluminium–magnesium alloys

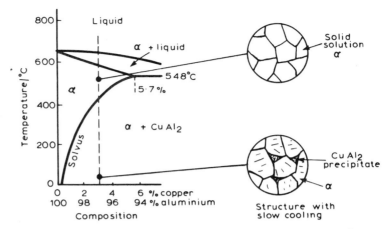

Figure 5.2 Thermal equilibrium diagram for aluminium–copper alloys

phase. When the temperature falls below the solvus temperature a copper–aluminium compound is precipitated. The result at room temperature is α solid solution with this copper–aluminium compound precipitate ($CuAl_2$). The precipitate is rather coarse, but this structure of the alloy can be changed by heating to about 500°C, soaking at that temperature, and then quenching, to give a supersaturated solid solution, just α phase with no precipitate. This treatment, known as *solution treatment*, results in an unstable situation. With time a fine precipitate will be produced. Heating to, say, 165°C for about ten hours hastens the production of this fine precipitate (*Figure 5.3*). The microstructure with this fine precipitate is both stronger and harder. The treatment is referred to as *precipitation hardening* (see later in this chapter).

A common group of heat treatable wrought alloys is based on aluminium with copper. Thus one form has 4.0% copper, 0.8% magnesium, 0.5% silicon and 0.7% manganese. This alloy is known as Duralumin. The heat treatment process used is solution treatment at 480°C, quenching and then precipitation hardening at either room temperature for about 4 days or 10 hours at 165°C. This alloy is widely used in aircraft bodywork. The presence of the copper does

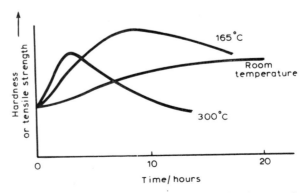

Figure 5.3 The effects of time and temperature on hardness and strength for an aluminium–copper alloy

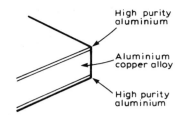

Figure 5.4 A clad duraluminium sheet

however reduce the corrosion resistance and thus the alloy is often clad with a thin layer of high purity aluminium to improve the corrosion resistance (*Figure 5.4*).

The precipitation hardening of an aluminium–copper alloy is due to the precipitate of the aluminium–copper compound. The age hardening of aluminium–copper–magnesium–silicon alloys is due to the precipitates of both an aluminium–copper compound $CuAl_2$ and an aluminium–copper–magnesium compound $CuAl_2Mg$. Other heat treatable wrought alloys are based on aluminium with magnesium and silicon. The age hardening with this alloy is due to the precipitate of a magnesium–silicon compound Mg_2Si. A typical alloy has the composition 0.7% magnesium, 1.0% silicon and 0.6% manganese. This alloy is not as strong as the duralumin but has greater ductility. It is used for ladders, scaffold tubes, container bodies, structural members for road and rail vehicles. The heat treatment is solution treatment at 510°C with precipitation hardening by quenching followed by precipitation hardening of 10 hours at about 165°C. Another group of alloys is based on aluminium–zinc–magnesium–copper, e.g. 5.5% zinc, 2.8% magnesium, 0.45% copper, 0.5% manganese. These alloys have the highest strength of the aluminium alloys and are used for structural applications in aircraft and spacecraft.

An alloy for use in the casting process must flow readily to all parts of the mould and on solidifying it should not shrink too much and any shrinking should not result in fractures. In choosing an alloy for casting, the type of casting process being used needs to be taken into account. In sand casting the mould is made of sand bonded with clay or a resin. The cooling rate with such a method is relatively slow. A material for use by this method must give a suitable strength material after a slow cooling process. With die casting the mould is made of metal and the hot metal is injected into the die under pressure. This results in a fast cooling. A material for use by this method must develop suitable strength after fast cooling.

A family of aluminium alloys that can be used in the 'as cast' condition, i.e. no heat treatment is used, has aluminium with between 9 and 13% silicon. These alloys can be used for both sand and die casting. The addition of the silicon increases the fluidity of the alloy.

The aluminium–silicon alloy is widely used for both sand and die casting, being used for many castings in cars, e.g. sumps, gear-boxes and radiators. It is also used for pump parts, motor housings and a wide variety of thin walled and complex castings.

Other cast alloys that are not heat treated are aluminium–silicon–copper alloys, e.g. 5.0% silicon and 3.0% copper, and aluminium–magnesium–manganese alloys, e.g. 4.5% magnesium and 0.5% manganese. The silicon–copper alloys can be both sand and die cast, the magnesium–manganese alloys are however only suitable for sand casting. They have excellent corrosion resistance and are often used in marine environments.

The addition of copper, magnesium and other elements to aluminium alloys, either singly or in some suitable combination, can enable the alloy to be heat treated. Thus an alloy having 5.5%

silicon and 0.6% magnesium can be subjected to solution treatment followed by precipitation hardening to give a high-strength casting material. Another heat treatable casting alloy has 4.0% copper, 2.0% nickel and 1.5% magnesium.

Example
Which of the following aluminium wrought alloys would be most suitable for use in a marine environment? (a) 1.25% manganese (b) 5.0% magnesium (c) 1.2% magnesium.

The aluminium alloy with the highest percentage magnesium (b) has the greatest corrosion resistance and so is most suitable for a marine environment.

PROPERTIES OF ALUMINIUM ALLOYS

The table shows the properties that might be obtained with aluminium alloys.

Composition	Condition	Tensile strength /N mm^{-2} (MPa)	Elongation /%	Hardness /HB
Wrought, non-heat treated alloys				
1.25% Mn	Annealed	110	30	30
	Hard	180	3	50
2.25% Mg	Annealed	180	22	45
	¾ hard	250	4	70
5.0% Mg	Annealed	300	16	65
	¼ hard	340	8	80
Wrought, heated treated alloys				
4.0% Cu, 0.8% Mg, 0.5% Si, 0.7% Mn	Annealed	180	20	45
	Solution treated, precipitation hardened	430	20	100
4.3% Cu, 0.6% Mg, 0.8% Si, 0.75% Mn	Annealed	190	12	45
	Solution treated, precipitation hardened	450	10	125
0.7% Mg, 1.0% Si, 0.6% Mn	Annealed	120	15	47
	Solution treated, precipitation hardened	300	12	100
5.5% Zn, 2.8% Mg, 0.45% Cu, 0.5% Mn	Solution treated, precipitation hardened	500	6	170
Cast, non-heat treated alloys				
12% Si	Sand cast	160	5	55
	Die cast	185	7	60
5% Si, 3% Cu	Sand cast	150	2	70
	Die cast	170	3	80
4.5% Mg, 0.5% Mn	Sand cast	140	3	60
Cast, heat treated alloys				
5.5% Si, 0.6% Mg	Sand cast, solution treated, precipitation hardened	235	2	85
4.0% Cu, 2% Ni, 1.5% Mg	Sand cast, solution treated, precipitation hardened	275	1	110

The effects of the various alloying elements used with aluminium can be summarised as:

Copper	Increases strength. Precipitation heat treatment possible. Improves machineability
Manganese	Improves ductility. Improves, in combination with iron, the castability
Magnesium	Improves strength. Precipitation heat treatment possible with more than about 6%. Improves the corrosion resistance
Silicon	Improves castability, giving an excellent casting alloy. Improves corrosion resistance
Zinc	Lowers castability. Improves strength when combined with other alloying elements

COPPER

Copper has a density of 8.93×10^3 kg m^{-3}. It has very high electrical and thermal conductivity and can be manipulated readily by either hot or cold working. Pure copper is very ductile and relatively weak. The tensile strength and hardness can be increased by working, this does however decrease the ductility. Copper has good corrosion resistance. This is because there is a surface reaction between copper and the oxygen in the air which results in the formation of a thin protective oxide layer.

Very pure copper can be produced by an electrolytic refining process. An impure slab of copper is used as the anode while a pure thin sheet of copper is used as the cathode. The two electrodes are suspended in a warm solution of dilute sulphuric acid (*Figure 5.5*). The passage of an electric current through the arrangement causes copper to leave the anode and become deposited on the cathode. The result is a thicker, pure copper cathode, while the anode effectively disappears, the impurities having fallen to the bottom of the container. The copper produced by this process is often called *cathode copper* and has a purity greater than 99.99%. It is used mainly as the raw material for the production of alloys, though there is some use as a casting material.

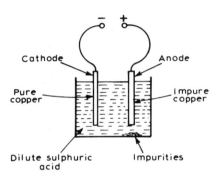

Figure 5.5 Basic arrangement for the electrolytic refining of copper

Electrolytic tough pitch high conductivity copper is produced from cathode copper which has been melted and cast into billets, and other suitable shapes, for working. It contains a small amount of oxygen, present in the form of cuprous oxide, which has little effect on the electrical conductivity of the copper. This type of copper should not be heated in an atmosphere where it can combine with hydrogen because the hydrogen can diffuse into the metal and combine with the cuprous oxide to generate steam. This steam can exert sufficient pressure to cause cracking of the copper.

Fire refined tough pitch high conductivity copper is produced from impure copper. In the fire refining process, the impure copper is melted in an oxidising atmosphere. The impurities react with the oxygen to give a slag which is removed. The remaining oxygen is partially removed by poles of green hardwood being thrust into the liquid metal, the resulting combustion removes oxygen from the metal. The resulting copper has an electrical conductivity almost as good as the electrolytic tough pitch high conductivity copper.

Oxygen-free high conductivity copper can be produced if, when cathode copper is melted and cast into billets, there is no oxygen present in the atmosphere. Such copper can be used in atmospheres where hydrogen is present.

Another method of producing oxygen-free copper is to add phosphorus during the refining. The effect of small amounts of phosphorus in the copper is a very marked decrease in the electrical conductivity, of the order of 20%. Such copper is known as *phosphorus deoxidised copper* and it can give good welds, unlike the other forms of copper.

The addition of about 0.5% arsenic to copper increases its tensile strength, especially at temperatures of about 400°C. It also improves its corrosion resistance but greatly reduces the electrical and thermal conductivities. This type of copper is known as *arsenical copper*.

Electrolytic tough pitch, high-conductivity copper finds use in high-grade electrical applications, e.g. wiring and busbars. Fire-refined, tough pitch, high-conductivity copper is used for standard electrical applications. Tough pitch copper is also used for heat exchangers and chemical plant. Oxygen-free, high-conductivity copper is used for high-conductivity applications where hydrogen may be present, electronic components and as the anodes in the electrolytic refining of copper. Phosphorus deoxidised copper is used in chemical plant where good weldability is necessary and for plumbing and general pipework. Arsenical copper is used for general engineering work, being useful to temperatures of the order of 400°C.

The table shows the properties that might be obtained with the various forms of copper.

Composition %	Condition	Tensile strength /N mm^{-2} (MPa)	Elongation %	Hardness /HB
Electrolytic tough-pitch, high-conductivity copper				
99.90 min 0.05 oxygen	Annealed	220	50	45
	Hard	400	4	115
Fire refined tough-pitch, high conductivity copper				
99.85 min 0.05 oxygen	Annealed	220	50	45
	Hard	400	4	115
Oxygen-free high-conductivity copper				
99.95 min	Annealed	220	60	45
	Hard	400	6	115
Phosphorus deoxidised copper				
99.85 min 0.013–0.05 P	Annealed	220	60	45
	Hard	400	4	115
Arsenical copper				
99.20 min 0.05 oxygen, 0.3–0.5 As	Annealed	220	50	45
	Hard	400	4	115

COPPER ALLOYS

The most common elements with which copper is alloyed are zinc, tin, aluminium and nickel. The copper–zinc alloys are referred to as brasses, the copper–tin alloys as tin bronzes, the copper–aluminium alloys as aluminium bronzes and the copper–nickel alloys as cupronickels, though where zinc is also present they are called nickel silvers. A less common copper alloy involves copper and beryllium.

The copper–nickel thermal equilibrium diagram is rather simple as the two metals are completely soluble in each other in both the liquid and solid states. The copper–zinc, copper–tin and copper–aluminium thermal equilibrium diagrams are however rather complex. In all cases, the α phase solid solutions have the same types of microstructure and are ductile and suitable for cold working. When the amount of zinc, tin or aluminium exceeds that required to saturate the α solid solution a β phase is produced. The microstructures of these β phases are similar and alloys containing this phase are stronger and less ductile. They cannot be readily cold worked and are hot worked or cast. Further additions lead to yet further phases which are hard and brittle.

The *brasses* are copper–zinc alloys containing up to about 43% zinc. *Figure 5.6* shows the relevant part of the thermal equilibrium diagram. Brasses with between 0 and 35% zinc solidify as α solid solutions. These brasses have high ductility and can readily be cold worked. *Gilding brass*, 15% zinc, is used for jewellery because it has a colour resembling that of gold and can so easily be worked. *Cartridge brass*, 30% zinc and frequently referred to as *70/30 brass*, is used where high ductility is required with relatively high strength. It is called cartridge brass because of its use in the production of cartridge and shell cases. The brasses in the 0 to 30% zinc range all have their tensile strength and hardness increased by working, but the ductilities decrease.

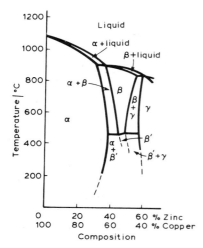

Figure 5.6 Thermal equilibrium diagram for copper–zinc alloys

Brasses with between 35 and 46% zinc solidify as a mixture of two phases. Between about 900°C and 453°C the two phases are α and β. At 453°C this β phase transforms to a low temperature modification referred to as β′ phase. Thus at room temperature the two phases present are α and β′. The presence of the β′ phase produces a drop in ductility but an increase in tensile strength to the maximum value for a brass (*Figure 5.7*). These brasses are known as *alpha-beta* or *duplex brasses*. These brasses are not cold worked but have good properties for hot forming processes, e.g. extrusion. This is because the β phase is more ductile than the β′ phase and hence the combination of α plus β gives a very ductile material. The hot working should take place at temperatures in excess of 453°C. The name *Muntz metal* is given to a brass with 60% copper–40% zinc.

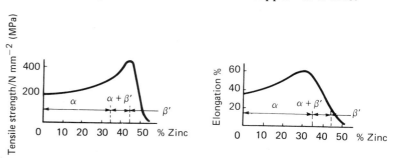

Figure 5.7 Strength and ductility for copper–zinc alloys

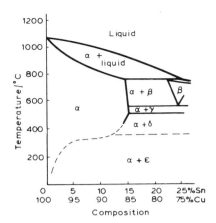

Figure 5.8 Thermal equilibrium diagram for tin bronzes

The addition of lead to Muntz metal improves considerably the machining properties, without significantly changing the strength and ductility. *Leaded Muntz metal* has 60% copper, 0.3 to 0.8% lead and the remainder zinc.

Copper–zinc alloys containing just the β' phase have little industrial application. The presence of γ phase in a brass results in a considerable drop in strength and ductility, a weak brittle product being obtained.

Copper–tin alloys are known as *tin bronzes*. *Figure 5.8* shows the thermal equilibrium diagram for such alloys. The dashed lines on the diagram indicate the phase that can occur with extremely slow cooling. The structure that normally occurs with up to about 10% tin are predominantly α solid solutions. Higher percentage tin alloys will invariably include a significant amount of the δ phase. This is a brittle intermetallic compound, the α phase being ductile.

Bronzes that contain up to about 8% tin are α bronzes and can be cold worked. In making bronze, oxygen can react with the metals and lead to a weak alloy. Phosphorus is normally added to the liquid metals to act as a deoxidiser. Some of the phosphorus remains in the final alloy. This type of alloy is known as *phosphor bronze*. These alloys are used for springs, bellows, electrical contacts, clips, instrument components. A typical phosphor bronze might have about 95% copper, 5% tin and 0.02 to 0.40% phosphorus.

The above discussion refers to wrought phosphor bronzes. Cast phosphor bronzes contain between 5 and 13% tin with as much as 0.5% phosphorus. A typical cast phosphor bronze used for the production of bearings and high grade gears has about 90% copper, 10% tin and a maximum of 0.5% phosphorus. This material is particularly useful for bearing surfaces; it has a low coefficient of friction and can withstand heavy loads. The hardness of the material occurs by virtue of the presence of both δ phase and a copper–phosphorus compound.

Casting bronzes that contain zinc are called *gunmetals*. This reduces the cost of the alloy and also makes unnecessary the use of phosphorus for deoxidation as this function is performed by the zinc. *Admiralty gunmetal* contains 88% copper, 10% tin and 2% zinc. This alloy finds general use for marine components, hence the word 'Admiralty'.

Copper–aluminium alloys are known as *aluminium bronzes*. *Figure 5.9* shows the thermal equilibrium diagram for such alloys. Up to about 9% aluminium gives *alpha bronzes*, such alloys containing just the α phase. Alloys with up to about 7% aluminium can be cold worked readily. *Duplex alloys* with about 10% aluminium are used for casting. Aluminium bronzes have high strength, good resistance to corrosion and wear. These corrosion and wear properties arise because of the thin film of aluminium oxide formed on the surfaces. Typical applications of such materials are high-strength and highly corrosion-resistant items in marine and chemical environments, e.g. pump casings, gears, valve parts.

Alloys of copper and nickel are known as *cupronickels*, though if zinc is also present they are referred to as *nickel silvers*. *Figure 3.8* shows the thermal equilibrium diagram for copper–nickel alloys.

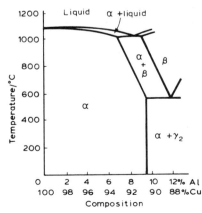

Figure 5.9 Thermal equilibrium diagram for aluminium bronzes

Copper and nickel are soluble in each other in both the liquid and solid states, they thus form a solid solution whatever the proportions of the two elements. They are thus α phase over the entire range and suitable for both hot and cold working over the entire range. The alloys have high strength and ductility, and good corrosion resistance. The 'silver' coinage in use in Britain is a 75% copper – 25% nickel alloy. The addition of 1 to 2% iron to the alloys increases their corrosion resistance. A 66% copper – 30% nickel – 2% manganese – 2% iron alloy is particularly resistant to corrosion, and erosion, and is used for components immersed in moving sea water.

Nickel silvers have a silvery appearance and find use for items such as knives, forks and spoons. The alloys can be cold worked and usually contain about 20% nickel, 60% copper and 20% zinc.

Copper alloyed with small percentages of beryllium can be precipitation heat treated to give alloys with very high tensile strengths, such alloys being known as *beryllium bronzes*, or *beryllium copper*. The alloys are used for high-conductivity, high-strength electrical components, springs, clips and fastenings.

Example
Two brasses, an alpha and a duplex brass, are available. Which brass should be selected for cold working?
The alpha brass.

Example
Basis brass has the composition 63% copper – 37% zinc. What type of mechanical properties might be expected of such a brass?
Brasses with between 0 and about 35% zinc solidify as alpha brasses. Basis brass is thus likely to be predominantly alpha but with some beta phase. Because of the predominant amount of alpha brass the brass is likely to be reasonably ductile and capable of being cold worked. The presence of a small amount of beta phase will result in an increase in tensile strength. The brass is likely therefore to be a reasonable compromise, having a relatively high tensile strength and ductility. In fact, basis brass is a general purpose alloy, being widely used for general hardware.

PROPERTIES OF COPPER

The table shows the properties that might be obtained with copper alloys.

Composition %	Condition	Tensile strength /N mm^{-2} (MPa)	Elongation %	Hardness /HB
Brasses				
90 copper, 10 zinc	Annealed	280	48	65
80 copper, 20 zinc	Annealed	320	50	67
70 copper, 30 zinc	Annealed	330	70	65
	Hard	690	5	185
60 copper, 40 zinc	Annealed	380	40	75

Non-ferrous alloys

Composition %	Condition	Tensile strength /N mm^{-2} (MPa)	Elongation %	Hardness /HB
Tin bronzes				
95 copper, 5 tin, 0.02–0.40 phosphorus	Annealed	340	55	80
	Hard	700	6	200
91 copper 8–9 tin, 0.02–0.40 phosphorus	Annealed	420	65	90
	Hard	850	4	250
Gunmetal				
88 copper, 10 tin, 2 zinc	Sand cast	300	20	80
Aluminium bronzes				
95 copper, 5 aluminium	Annealed	370	65	90
	Hard	650	15	190
88 copper, 9.5 aluminium, 2.5 iron	Sand cast	545	30	110
Cupronickels				
87.5 copper, 10 nickel, 1.5 iron, 1 manganese	Annealed	320	40	155
75 copper, 25 nickel, 0.5 manganese	Annealed	360	40	90
	Hard	600	5	170
Nickel silver				
64 copper, 21 zinc, 15 nickel	Annealed	400	50	100
	Hard	600	10	180
Beryllium bronzes				
98 copper, 1.7 beryllium, 0.2 to 0.6 cobalt and nickel	Solution treated, precipitation hardened	1200	3	370

There are a considerable number of copper alloys. The following indicates the type of selection that could be made.

Electrical conductors	Electrolytic tough-pitch, high-conductivity copper
Tubing and heat exchangers	Phosphorus deoxidised copper is generally used. Muntz metal, cupronickels or naval brass (62% copper – 37% zinc – 1% tin) is used if the water velocities are high
Pressure vessels	Phosphorus deoxidised copper, copper-clad steel or aluminium bronze
Bearings	Phosphor bronze. Other bronzes and brasses with some lead content are used in some circumstances
Gears	Phosphor bronze. For light duty gunmetals, aluminium bronze or die cast brasses may be used
Valves	Aluminium bronze
Springs	Phosphor bronze, nickel silver, basis brass are used for low cost springs. Beryllium bronze is the best material

MAGNESIUM

Magnesium has a density of 1.7×10^3 kg m^{-3} and thus a very low density compared with other metals. It has an electrical conductivity of about 60% of that of copper, as well as a high thermal conductivity. It has a low tensile strength, needing to be alloyed with other metals to improve its strength. Under ordinary atmospheric conditions magnesium has good corrosion resistance, which is provided by an oxide layer that develops on the surface of the magnesium in air. However, this oxide layer is not completely impervious, particularly in air that contains salts, and thus the corrosion resistance can be low under adverse conditions. Magnesium is only used generally in its alloy form, the pure metal finding little application.

MAGNESIUM ALLOYS

Because of the low density of magnesium, the magnesium-base alloys have low densities. Thus magnesium alloys are used in applications where lightness is the primary consideration, e.g. in aircraft and spacecraft. Aluminium alloys have higher densities than magnesium alloys but can have greater strength. The strength-to-weight ratio for magnesium alloys is however greater than that for aluminium alloys. Magnesium alloys also have the advantage of good machineability and weld readily.

Magnesium–aluminium–zinc alloys and magnesium–zinc–zirconium are the main two groups of alloys in general use. Small amounts of other elements are also present in these alloys. The composition of an alloy depends on whether it is to be used for casting or working, i.e. a wrought alloy. The cast alloys can often be heat treated to improve their properties.

A general-purpose wrought alloy has about 93% magnesium – 6% aluminium – 1% zinc – 0.3% manganese. This alloy can be forged, extruded and welded, and has excellent machinability. A high strength wrought alloy has 96.4% magnesium – 3% zinc – 0.6% zirconium. A general-purpose casting alloy has about 91% magnesium – 8% aluminium – 0.5% zinc – 0.3% manganese. A high-strength casting alloy has 94.8% magnesium – 4.5% zinc – 0.7% zirconium. Both these casting alloys can be heat treated.

The table shows the properties that might be obtained with magnesium alloys.

Composition %	Condition	Tensile strength /N mm^{-2} (MPa)	Elongation %	Hardness /HB
Wrought alloys				
93 magnesium, 6 aluminium, 1 zinc, 0.3 manganese	Forged	290	8	65
	Extruded	215	8	
96.4 magnesium, 3 zinc, 0.6 zirconium	Rolled	265	8	70
	Extruded	310	8	
Cast alloys				
91 magnesium, 8 aluminium, 0.5 zinc, 0.6 zirconium	As cast	140	2	55
	Heat treated	200	6	75
94.8 magnesium, 4.5 zinc, 0.7 zirconium	Heat treated	230	5	70

NICKEL

Nickel has a density of 8.88×10^3 kg m^{-3} and a melting point of 1455°C. It possesses excellent corrosion resistance, hence it is used often as a cladding on a steel base. This combination allows the corrosion resistance of the nickel to be realised without the high cost involved in using entirely nickel. Nickel has good tensile strength and maintains it at quite elevated temperatures. Nickel can be both cold and hot worked, has good machining properties and can be joined by welding, brazing and soldering.

Nickel is used in the food processing industry, in chemical plant, and in the petroleum industry, because of its corrosion resistance and strength. It is also used in the production of chromium-plated mild steel, the nickel forming an intermediate layer between the steel and the chromium. The nickel is electroplated on to the steel.

NICKEL ALLOYS

Nickel is used as the base metal for a number of alloys with excellent corrosion resistance and strength at high temperatures. One group of alloys is based on nickel combined with copper, the thermal equilibirum diagram for these alloys being on page 45. A common nickel–copper alloy is known as *Monel* which has 68% nickel, 30% copper and 2% iron. It is highly resistant to sea water, alkalis, many acids and superheated steam. It has also high strength, hence its use for steam turbine blades, food processing equipment and chemical engineering plant components.

Another common nickel alloy is known as *Inconel* which contains 78% nickel, 15% chromium and 7% iron. The alloy has a high strength and excellent resistance to corrosion at both normal and high temperatures. It is used in chemical plant, aero-engines, as sheaths for electric cooker elements, steam turbine parts and heat treatment equipment.

The *Nimonic* series of alloys are basically nickel–chromium alloys, essentially about 80% nickel and 20% chromium. They have high strength at high temperatures and are used in gas turbines for discs and blades.

The table shows the properties that might be obtained with nickel alloys.

Composition %	Condition	Tensile strength /N mm^{-2} (MPa)	Elongation %	Hardness /HB
68 nickel, 30 copper, 2 iron	Annealed	500	40	110
	Cold worked	840	8	240
78 nickel, 15 chromium, 7 iron	Annealed	700	35	170
	Cold worked	1050	15	290

TITANIUM

Titanium has a relatively low density, 4.5×10^3 kg m^{-3}, just over half that of steel. It has a relatively low tensile strength when pure but alloying gives a considerable increase in strength. Because of the low density of titanium its alloys have a high strength-to-weight ratio. Also, it has excellent corrosion resistance. However, titanium is an expensive metal, its high cost reflecting the difficulties experienced in the extraction and forming of the material; the ores are quite plentiful.

TITANIUM ALLOYS

The main alloying elements used with titanium are aluminium, copper, manganese, molybdenum, tin, vanadium and zirconium. A ductile, heat treatable alloy has 97.5% titanium – 2.5% copper and it can be welded and formed. A higher-strength alloy, which can also be welded and formed, has 92.5% titanium – 5% aluminium – 2.5% tin. A very high strength alloy has 82.5 titanium – 11% tin – 4% molybdenum – 2.25% aluminium – 0.25% silicon. This alloy can be heat-treated and forged.

The titanium alloys all show excellent corrosion resistance, have good strength-to-weight ratios, can have high strengths, and have good properties at high temperatures. They are used for compressor blades, engine forgings, components in chemical plant, and other duties where their properties make them one of the few possible choices despite their high cost.

The table shows the properties that might be obtained with titanium alloys.

Composition %	Condition	Tensile strength /N mm^{-2} (MPa)	Elongation %	Hardness /HB
97.5 titanium, 2.5 copper	Heat treated	740	15	360
92.5 titanium, 5 aluminium, 2.5 tin	Annealed	880	16	360
82.5 titanium, 11 tin, 4 molybdenum, 2.25 aluminium, 0.25 silicon	Heat treated	1300	15	380

ZINC

Zinc has a density of 7.1×10^3 kg m^{-3}. Pure zinc has a melting point of only 419°C and is a relatively weak metal. It has good corrosion resistance, due to the formation of an impervious oxide layer on the surface. Zinc is frequently used as a coating on steel in order to protect that material against corrosion, the product being known as galvanised steel.

ZINC ALLOYS

The main use of zinc alloys is for die-casting. They are excellent for this purpose by virtue of their low melting points and the lack of corrosion of dies used with them. The two alloys in common use for this purpose are known as alloy A and alloy B. *Alloy A*, the most used of the two, has the composition of 3.9 to 4.3% (max) aluminium, 0.03% (max) copper, 0.03 to 0.06% (max) magnesium, the remainder being zinc. *Alloy B* has the composition 3.9 to 4.3% (max) aluminium, 0.75 to 1.25% (max) copper, 0.03 to 0.06% (max) magnesium, with the remainder being zinc. Alloy A is the more ductile, alloy B has the greater strength.

The zinc used in the alloys has to be extremely pure so that little, if any, other impurities are introduced into the alloys, typically the required purity is 99.99%. The reason for this purity is that the presence of very small amounts of cadmium, lead or tin renders the alloy susceptible to intercrystalline corrosion. The products of this corrosion cause a casting to swell and may lead to failure in service.

After casting, the alloys undergo a shrinkage which takes about a month to complete; after that there is a slight expansion. A casting can be *stabilised* by annealing at 100°C for about 6 hours.

Zinc alloys can be machined and, to a limited extent, worked. Soldering and welding are not generally feasible.

Zinc alloy die-castings are widely used in domestic appliances, for toys, car parts such as door handles and fuel pump bodies, optical instrument cases.

The table shows the properties that might be obtained with zinc diecasting alloys.

Composition %	Condition	Tensile strength /N mm^{-2} (MPa)	*Elongation* %	*Hardness* /HB
Alloy A	As cast	285	10	83
Alloy B	As cast	330	7	92

COMPARISON OF NON-FERROUS ALLOYS

The following table shows the range of strengths that are obtained with the non-ferrous alloys considered in this chapter.

Alloys	*Tensile strength*/N mm^{-2} (MPa)
Aluminium	100 to 550
Copper	200 to 1300
Magnesium	150 to 350
Nickel	400 to 1300
Titanium	400 to 1600
Zinc	200 to 350

Thus, if the main consideration in the choice of an alloy is that it must have high strength, then the choice, if limited to the alloys listed above, would be titanium, with the second choice nickel or copper alloys.

In some applications it is not just the strength of a material that is important but the strength/weight ratio. This is particularly the case in aircraft or spacecraft where not only is strength required but also a low mass. The following table shows the densities of the alloys and the tensile strength/density ratio.

Alloys	*Density* /10^3 kg m^{-3}	*Tensile strength/density* /N mm^{-2} × 10^3 kg m^{-3}
Aluminium	2.7	37 to 200
Copper	8	25 to 160
Magnesium	1.8	110 to 190
Nickel	8.9	47 to 146
Titanium	4.5	89 to 356
Zinc	6.7	30 to 52

Titanium gives the best possible strength/weight ratio. Magnesium, with its relatively low strength, has however a fairly high strength/weight ratio by virtue of its very low density. Aluminium, magnesium and titanium alloys are widely used in aircraft.

Another consideration that may affect the choice of an alloy is its corrosion resistance. In general the most resistant are titanium alloys, with the least resistant being magnesium. The rough order of descending corrosion resistance is: titanium, copper, nickel, zinc, aluminium and magnesium.

A vital factor in considering the choice of a material is the cost. The following table gives the relative costs of the alloys, the costs all being relative to aluminium. The costs are given in terms of the cost per unit mass.

Alloy	Relative cost per unit mass
Aluminium	1
Copper	1 to 2
Magnesium	3
Nickel	5
Titanium	25
Zinc	0.5

Titanium is the most expensive of the alloys, zinc the least expensive. However in practice the use of a high-strength material may mean that a thinner section can be used and so less mass of alloy is required.

The process to be used to shape the material is another constraint on the selection of the material. With ductile materials, the drag forces on a cutting tool are high and the swarf tends to build up. These lead to poor machinability. Annealed aluminium has a high ductility and thus has poor machinability. A half-hard aluminium alloy can however have a good machinability, the ductility being much less. Materials can have their machinability improved by introducing other materials, such as lead, into the alloy concerned. The introduced material is in the form of particles which help to break up the swarf into small chips. Very hard materials may present machining problems in that the cutting tool needs to be harder than the material being machined.

Example
Which type of alloy would be the optimum choice for a new high-speed aircraft where high strength at high temperatures combined with a low density is required?

Titanium alloys give high strengths at high temperatures and have a low density.

Example
Which type of alloy combines the following properties: can be die-cast, reasonable corrosion resistance and cheap? Strength is not particularly required.

Zinc alloys can be die-cast and are cheap. They also have reasonable corrosion resistance.

PRECIPITATION HARDENING

If a solution of sodium chloride in water is cooled sufficiently, sodium chloride precipitates out of the solution. This occurs because the solubility of sodium chloride in water decreases as the temperature decreases. Thus very hot water may contain 37 g per 100 g of water. Cold water is saturated with about 36 g. When the hot solution cools down the surplus salt is precipitated out of the solution. Similar events can occur with solid solutions.

Figure 5.10 shows part of the copper–silver thermal equilibrium diagram. When the 5% copper – 95% silver alloy is cooled from the liquid state to 800°C a solid solution is produced. At this temperature the solid solution is not saturated but, cooling to the solvus temperature makes the solid solution saturated. If the cooling is continued, slowly, a precipitate occurs. The result at room temperature is a solid solution in which a coarse precipitate occurs.

The above discussion assumes that the cooling occurs very slowly. The formation of a precipitate requires the grouping together of atoms. This requires atoms to diffuse through the solid solution. Diffusion is a slow process, if the solid solution is cooled rapidly from 800°C, i.e. quenched, the precipitation may not occur. The solution becomes *supersaturated*, i.e. it contains more of the α phase than the equilibrium diagram predicts. The result of this rapid cooling is a solid solution, the α phase, at room temperature.

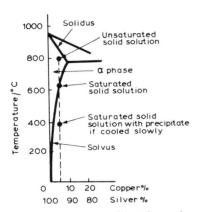

Figure 5.10 Copper–silver thermal equilibrium diagram

The supersaturated solid solution may be retained in this form at room temperature, but the situation is not very stable and a very fine precipitation may occur with the elapse of time. This precipitation may be increased if the solid is heated for some time (the temperature being significantly below the solvus temperature). The precipitate tends to be very minute particles dispersed throughout the solid. Such a fine dispersion gives a much stronger and harder alloy than when the alloy is cooled slowly from the α solid solution. This hardening process is called *precipitation hardening*. The term *natural ageing* is sometimes used for the hardening process that occurs due to precipitation at room temperature and the term *artificial ageing* when the precipitation occurs as a result of heating.

Figure 5.11 shows part of the thermal equilibrium diagram for aluminium–copper alloys. If the alloy with 4% copper is heated to

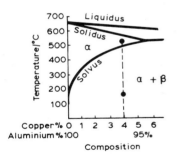

Figure 5.11 Aluminium–copper thermal equilibrium diagram

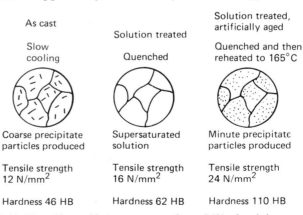

Figure 5.12 The effect of heat treatment for a 96% aluminium – 4% copper alloy

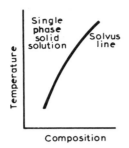

Figure 5.13 The required form of solvus line for precipitation hardening

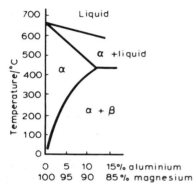

Figure 5.14 Thermal equilibrium diagram for aluminium–magnesium alloys

about 500°C and held at that temperature for a while, diffusion will occur and a homogeneous α solid solution will form. If the alloy is then quenched to about room temperature supersaturation occurs. This quenched alloy is relatively soft. If now the alloy is heated to a temperature of about 165°C and held at this temperature for about ten hours a fine precipitate is formed. *Figure 5.12* shows the effects on the alloy structure and properties of these processes. The effect is to give an alloy with a higher tensile strength and harder.

Not all alloys can be treated in this way. Precipitation hardening can only occur, in a two-metal alloy, if one of the alloying elements has a high solubility at high temperatures and a low solubility at low temperatures, i.e. the solubility decreases as the temperature decreases. This means that the solvus line must slope as shown in *Figure 5.13*. Also the structure of the alloy at temperatures above the solvus line must be a single phase solid solution. The alloy systems that have some alloy compositions that can be treated in this way are mainly non-ferrous, e.g. copper–aluminium and magnesium–aluminium.

Example
Which of the following alloys would you anticipate could be precipitation hardened?
(a) 95% magnesium – 5% aluminium alloy (see *Figure 5.14*).
(b) 85% magnesium – 15% aluminium alloy (see *Figure 5.14*).
(a) would be expected to be capable of being precipitation hardened on the basis of the slope of the solvus; (b) has a composition that has too much aluminium.

PROBLEMS

1 What is the effect on the strength and ductility of aluminium of (a) the purity of the aluminium, (b) the temper?

2 Describe the effect on the strength of aluminium of the percentage of magnesium alloyed with it.

3 Describe the solution treatment and precipitation hardening processes for aluminium–copper alloys.

4 Describe the features of aluminium–silicon alloys which makes them suitable for use with die-casting.

5 What is the effect of heat treatment on the properties of an aluminium alloy, such as a 4.3% copper – 0.6% magnesium – 0.8% silicon – 0.75% manganese?

6 Ladders are often made from an aluminium alloy. What are the properties required of that material in this particular use?

7 Explain with the aid of a thermal equilibrium diagram how the addition of a small amount of sodium to the melt of an aluminium–silicon alloy changes the properties of alloys with, say, 12% silicon.

8 What are the differences between the following forms of copper: electrolytic tough-pitch, high-conductivity copper; fire-refined, tough-pitch, high-conductivity copper; oxygen-free, high-conductivity copper; phosphorus-deoxidised copper; arsenical copper?

9 Which form of copper should be used in an atmosphere containing hydrogen?

10 What is the effect of cold work on the properties of copper?

11 What is the effect of the percentage of zinc in a copper–zinc alloy on its strength and ductility?

12 What brass composition would be most suitable for applications requiring (a) maximum tensile strength, (b) maximum ductility, (c) the best combined tensile strength and ductility?

13 In general, what are the differences in (a) composition, (b) properties of α phase and duplex alloys of copper?

14 The 'silver' coinage used in Britain is made from a cupro-nickel alloy. What are the properties required of this material for such a use?

15 What are the constituent elements in (a) brasses, (b) phosphor bronzes and (c) cupro-nickels?

16 The name Muntz metal is given to a 60% copper – 40% zinc alloy. Some forms of this alloy also include a small percentage of lead. What is the reason for the lead?

17 What percentage of aluminium would be likely to be present in an aluminium bronze that it to be cold worked?

18 What is meant by the term strength-to-weight ratio? Magnesium alloys have a high strength-to-weight ratio; of what significance is this in the uses to which the alloys of magnesium are put?

19 How does the corrosion resistance of magnesium alloys compare with other non-ferrous alloys?

20 What are the general characteristics of the nickel–copper alloys known as Monel?

21 Though titanium alloys are expensive compared with other non-ferrous alloys, they are used in modern aircraft such as *Concorde*. What advantages do such alloys possess which outweigh their cost?

22 What problems can arise when impurities are present in zinc die-casting alloys?

23 What is the purpose of 'stabilising annealing' for zinc die-casting alloys?

24 Describe the useful features of zinc die-casting alloys which makes them so widely used.

25 This question is based on the copper–zinc thermal equilibrium diagram given in *Figure 5.6*.

(a) What is the temperature at which a 85% copper – 15% zinc alloy begins to solidify?

(b) The pouring temperature used for casting is about 200°C above the liquidus temperature. What is the pouring temperature for the above alloy?

26 This question is based on the copper–zinc thermal equilibrium diagram given in *Figure 5.6*.

(a) Which of the following brasses would you expect to be just α solid solution? (i) 10% zinc – 90% copper, (ii) 20% zinc – 80% copper, (iii) 40% zinc – 60% copper.

(b) How would you expect the properties of the above brasses to differ? How are the differences related to the phases present in the alloys?

27 The following group of questions is concerned with justifying the choice of a particular alloy for a specific application.

(a) Why are zinc alloys used for die-casting?

(b) Why are magnesium alloys used in aircraft?

(c) Why are milk bottle caps made of an aluminium alloy?

(d) Why are titanium alloys extensively used in high-speed aircraft?

(e) Why are domestic water pipes made of copper?

(f) Why is cartridge brass used for the production of cartridge cases?

(g) Why are the ribs of hang gliders made of an aluminium alloy?

(h) Why are nickel alloys used for gas-turbine blades?

(i) Why is brass used for cylinder lock keys?

(j) Why are kitchen pans made of aluminium or copper alloys?

6 Polymeric materials

Objectives: At the end of this chapter you should be able to:

Describe the basic structures of polymers.
Explain how polymer crystallinity affects polymer properties.
Explain the significance of the glass transition temperature.
Explain the effects of temperature and time on the properties of a polymer.
Explain the effect of orientating polymer chains on the properties of the polymer.
Explain how the properties of polymers may be modified by additives.
Describe the benefits that can be obtained from copolymerisation.
Describe the properties and applications of common thermoplastic and thermosetting polymers.
Explain the characteristic structure of an elastomer and its mechanical properties.
Describe the properties and applications of common elastomers.
Describe the applications of polymer foams, polymer based sandwich materials and polymer composites.

POLYMERS

The term *polymer* is used to indicate that a compound consists of many repeating structural units. The prefix 'poly' means many. Each structural unit in the compound is called a *monomer*. Thus the plastic polyethylene is a polymer which has as its monomer the substance ethylene. For many plastics the monomer can be determined by deleting the prefix 'poly' from the name of the polymer.

If you apply heat to a plastic washing-up bowl the material softens. Removal of the heat causes the material to harden again. Such a material is said to be *thermoplastic*. The term implies that the material becomes 'plastic' when heat is applied.

If you applied heat to a plastic cup you might well find that the material did not soften but charred and decomposed. Such a material is said to be a *thermosetting plastic*.

Another type of polymer is the elastomer. Rubber is an elastomer. An *elastomer* is a polymer which by its structure allows considerable extensions which are reversible.

The thermoplastic, thermosetting and elastomer materials can be distinguished by their behaviour when forces are applied to them to cause stretching. Thermoplastic materials are generally flexible and relatively soft; if heated they become softer and more flexible. Thermosetting materials are rigid and hard with little change with an increase in temperature. Elastomers can be stretched to many times their initial length and still spring back to their original length when released. These different types of behaviours of polymers can

STRUCTURE OF POLYMERS

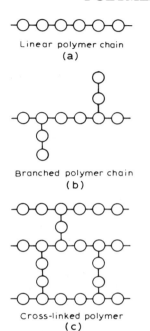

Figure 6.1 Forms of polymer structure

Figure 6.2 A linear amorphous polymer. Individual atoms are not shown, the chains being represented by lines

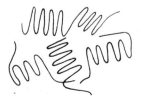

Figure 6.3 Folded linear polymer chains

Polymers involve the combining of many small molecules to give a large molecule involving very large numbers of atoms. The resulting large molecule may be in the form of a long linear chain, a chain with side branches or a cross-linked polymer giving rise to a network of interlocked atoms (*Figure 6.1*).

A crystal can be considered to be an orderly packing-together of atoms. The molecular chains of a polymer may however be completely tangled up in a solid with no order whatsoever (*Figure 6.2*). Such a material is said to be *amorphous*. There is however the possibility that the polymer molecules can be arranged in an orderly manner within a solid. Thus *Figure 6.3* shows linear polymer molecules folded to give regions of order. The orderly parts of such polymeric materials are said to be *crystalline*. Because long molecules can easily become tangled up with each other, polymeric materials are often only partially crystalline, i.e. parts of the material have orderly arrangements of molecules while other parts are disorderly.

Not all polymers can give rise to crystallinity. It is most likely to occur with simple linear chain molecules. Branched polymer chains are not easy to pack together in a regular manner, the branches get in the way. If the branches are completely regularly spaced along the chain then some crystallinity is possible, irregularly spaced branches render crystallinity improbable. Cross-linked polymers cannot be rearranged due to the links between chains and so crystallinity is not possible.

There are many polymers based on the form of the polyethylene molecule. Despite being linear molecules they do not always give rise to crystalline structures. PVC is essentially just the polyethylene molecule with some of the hydrogen atoms replaced with chlorine atoms (*Figure 6.4*). The molecule does not however give rise to crystalline structures. This is because the chlorine atoms are rather bulky and are not regularly spaced along the molecular chain. It is this lack of regularity which makes packing of the PVC molecular chains too difficult. Polypropylene has a molecule rather like that of polythene but with some of the hydrogen atoms replaced with CH_3

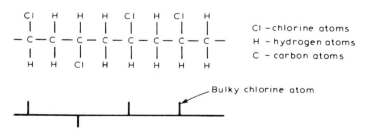

Figure 6.4 The basic form of a PVC molecule. The chlorine atoms are generally irregularly arranged on the different sides of the chain, so rendering orderly packing of the chains difficult

groups. These are however regularly spaced along the molecular chain and thus orderly packing is possible and so some degree of crystallinity.

The following table shows the form of the molecular chains and the degree of crystallinity possible for some common polymers.

Polymer	Form of chain	Possible crystallinity %
Polyethylene	Linear	95
	Branched	60
Polypropylene	Regularly spaced side groups on linear chain	60
Polyvinyl chloride	Irregularly spaced bulky chlorine atoms (*fig.* 6.4)	0
Polystyrene	Irregularly spaced bulky side groups	0

When an amorphous polymer is heated it shows no definite melting temperature but progressively becomes less rigid. The molecular arrangement in an amorphous material is all disorderly, just like that which occurs in a liquid. It is for this reason, i.e. no structural change occurring, that no sharp melting point occurs. For crystalline polymers there is an abrupt change at a particular temperature. Thus if the density of the polymer were being measured as a function of temperature, an abrupt change in density would occur at a particular temperature. At this temperature the crystallinity of the polymer disappears, the structure changing from a relatively orderly one below the temperature to a disorderly one above it. The temperature at which the crystallinity disappears is defined as being the *melting point* of the polymer. The following are the melting points of some common polymers.

Polymer	Crystallinity %	Melting point/°C
Polyethylene	95	138
Polyethylene	60	115
Polypropylene	60	176
Polyvinyl chloride	0	212*
Polystyrene	0	—

* This is not a clear melting point but a noticeable softening region of temperature.

The degree of crystallinity of a polymer affects its mechanical properties. The more crystalline a polymer the higher its tensile modulus. Thus the linear chain form of polyethylene in its crystalline form has the molecules closely packed together. Greater forces of attraction can exist between the chains when they are closely packed. The result is a stiffer material, a material with a higher tensile modulus. The branched form of polyethylene has a lower crystallinity and thus a lower tensile modulus. This is because the lower degree of crystallinity means that the molecules are not so closely packed together, an orderly structure is easier to pack closely than a disorderly one (you can get more clothes in a drawer if you pack them in an orderly manner than if you just throw them in). The further apart the molecular chains the lower the forces of attraction between them and so the less stiff, i.e. more flexible, the material and hence the lower the tensile modulus. The following are the tensile modulus and tensile strength values for polyethylene with different degrees of crystallinity.

Polymer	Crystallinity %	Tensile modulus /kN mm^{-2} (GPa)	Tensile strength /N mm^{-2} (MPa)
Polyethylene	95	21 to 38	0.4 to 1.3
Polyethylene	60	7 to 16	0.1 to 0.3

The more cross-links there are with a polymer the stiffer the polymer is, i.e. the higher its tensile modulus. A highly cross-linked polymer may be a rather hard, brittle substance.

GLASS TRANSITION TEMPERATURE

PVC, without any additives and at room temperature, is a rather rigid material. It is often used in place of glass. But if it is heated to a temperature of about 87°C a change occurs, the PVC becomes flexible and rubbery. The PVC below this temperature gives only a moderate elongation before breaking, above this temperature it stretches a considerable amount and behaves rather like a strip of rubber.

Polythene is a flexible material at room temperature and will give considerable extensions before breaking. If, however, the polythene is cooled to below about −120°C it becomes a rigid material.

The temperature at which a polymer changes from a rigid to a flexible material is called the *glass transition temperature*. The material is considered to be changing from a glass-like material to a rubber-like material. The following are the glass-transition temperatures for some common thermoplastics.

Material	Glass transition temperature/°C
Polyethylene	−120
Polypropylene	−10
Polyvinyl chloride	87
Polystyrene	100

At the glass transition temperature the tensile modulus shows an abrupt change from a high value for the glass-like material to a low value for the rubber-like material. *Figure 6.5* shows the form of the stress–strain graph for a polymer both below and above this transition temperature. Below the transition temperature the material has the type of stress–strain graph characteristic of a relatively brittle substance; above the transition temperature the graph is more like that of a rubber-like material in that the polymer may be stretched to many times its original length.

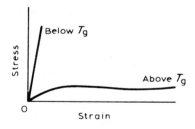

Figure 6.5 Stress–strain graphs for a polymer below and above its glass transition temperature T_g

Above the glass transition temperature the polymer chains are reasonably free to move and thus the application of stress causes molecular chains to uncoil and/or slide over one another, hence producing large extensions. When the temperature is decreased the density of the polymer increases and the chains become more closely packed. At the glass transition temperature the packing has become close enough to hinder the movement of the polymer chains to such an extent that the application of stress results in little extension.

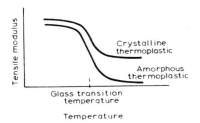

Figure 6.6 The effect of temperature on the tensile modulus

The value of the glass transition temperature varies from one polymer to another. It is lowest for those polymers that have linear molecules, e.g. polythene, which are flexible and easily packed together in the solid. Polymers having branched chains have higher glass transition temperatures, e.g. polyvinyl chloride, and are thus not easily packed, if at all, in an orderly manner.

Example

Should a polymer be below or above its glass transition temperature if it is to have (a) a high value of tensile modulus, (b) high elongations?

(a) Below the glass transition temperature (see *Figure 6.6*).

(b) Above the glass transition temperature, as it then behaves more like a rubber. Below the glass transition temperature the material is relatively brittle and thus has low elongations.

THE EFFECTS OF TEMPERATURE AND TIME ON MECHANICAL PROPERTIES

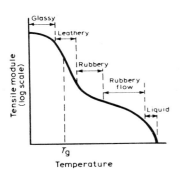

Figure 6.7 Variation of tensile modulus with temperature for an amorphous polymer

For polymers the variation of the tensile modulus with temperature is of considerable significance in that, for instance, it determines whether the plastic spoon used to stir the coffee is stiff and useful for the job concerned or far from stiff and not particularly useful. The tensile modulus of polymers decreases with an increase in temperature. The overall way in which the modulus changes with temperature is the same for all non-crystalline polymers, the differences between polymers being just the actual temperatures at which the changes occur. *Figure 6.7* shows the general type of variation that occurs with temperature – note the log scale for the modulus axis of the graph.

In the glassy range of the graph the tensile modulus is at its maximum value and is less dependent on temperature than at other temperatures. In this region the polymer is stiff and not so easily stretched. This is because the polymer chains are too closely packed to slide past each other and the only way the polymer can extend is by stretching the chains themselves.

The region between the glassy region and the rubbery region is referred to as the leathery region. This represents the transition of the polymer from a tightly packed structure where no chain movement was possible to a state where the polymer chains can uncoil from their tangled state, but spring back to their tangled state when the applied stress is removed.

At yet higher temperatures the chains begin to slip past each other when stress is applied. This is the rubbery flow region. At yet higher temperatures the polymer becomes liquid and complete movement of polymer chains past each other occurs. When chains move past each other the deformation is irrecoverable.

A plastic coffee spoon at room temperature is fairly rigid and if stressed may break in a brittle manner. The spoon is in the glassy region. If the spoon is in boiling water it behaves completely differently, very little stress has to be applied to cause a permanent irrecoverable deformation due to chain slipping occurring, the rubbery flow region.

The above discussion has considered what happens when stress is applied to polymers, but no mention was made of whether the stress

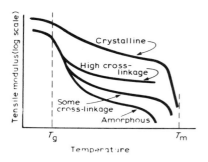

Figure 6.8 Variation of tensile modulus with temperature for polymers

was applied quickly or slowly. Time is needed for uncoiling or movement of polymer chains. A polymer that, at some particular temperature, may be rubbery with a slow application of stress, may be quite glass-like with a faster application of stress. The effect of increasing the rate of application of stress is to make the polymer more brittle and have a higher tensile modulus.

The effect of crystallinity on the behaviour of a polymer when subject to stress is to make it more stiff, i.e. have a higher tensile modulus. The polymer also becomes more brittle. *Figure 6.8* shows how the tensile modulus varies with temperature for a crystalline polymer.

The graph also shows how cross-links in a polymer affect its properties. With an increasing amount of cross-linkage so the rubbery and flow ranges disappear. With a high amount of cross-linkage the material is hard and fairly brittle.

TEMPERATURE AND POLYMER USE

Amorphous polymers tend to be used below their glass transition temperature T_g. They are however formed and shaped at temperatures above the glass transition temperature when they are in a soft condition. Crystalline polymers are used up to their melting temperature T_m. They can be hot-formed and shaped at temperatures above T_m, being cold-formed and shaped at temperatures between T_g and T_m.

Polythene is a crystalline polymer, with a melting point of 138°C and a glass transition temperature of −120°C for the form that gives 95% crystallinity. The maximum service temperature of polythene items is about 125°C, i.e. just below the melting point. The form of polythene with 60% crystallinity has a melting point of 115°C and a maximumn service temperature of about 95°C.

Polystyrene is an amorphous polymer with a glass transition temperature of 100°C. It has a maximum service temperature of about 80°C, i.e. just below the glass transition temperature.

Example
What would you expect to be the maximum service temperature for polyvinyl chloride? It is an amorphous polymer with a glass transition temperature of 87°C.

An amorphous polymer is used below its glass transition temperature and thus a maximum service temperature of about 70°C can be expected.

ORIENTATION

Figure 6.9 shows a typical stress–strain graph for a crystalline polymer, e.g. polythene. When the stress reaches point 'A' the material shows a sudden large reduction in cross-sectional area at some point (*Figure 6.10*). After this initial necking a considerable increase in strain takes place at essentially a constant stress, as the necked area gradually spreads along the entire length of the material. When the entire piece of material has reached the necked stage an increase in stress is needed to increase the strain further.

The above sequence of events can be explained by considering the orientation of the polymer chains. Prior to necking starting, the

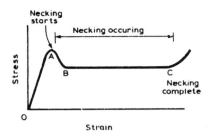

Figure 6.9 Typical stress–strain graph for a crystalline polymer

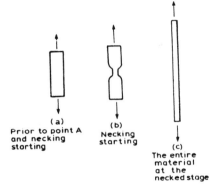

Figure 6.10 Necking in a polymer

polymer chains are folded to give regions of order in the material (as in *Figure 6.3*). When necking starts the polymer chains unfold to give a material with the chains lying along the direction of the forces stretching the material. As the necking spreads along the material so more of the chains unfold and line up. Eventually when the entire material is at the necked stage all the chains have lined up. The material in this state behaves differently to earlier in the stress–strain graph. The material is said to be *cold drawn*. It is completely crystalline, i.e. all the chains are packed in a very orderly manner.

The above sequence of events only tends to occur if the material is stretched slowly and sufficient time elapses for the molecular chains to slide past each other. If a high strain rate is used the material is likely to break without it becoming completely orientated. You can try pulling a strip of polythene for yourself and see the changes. (A strip cut from a polythene bag can be used and you can pull it in your hands.)

With crystalline or semi-crystalline polymers orientation is produced at temperatures both below and above the glass transition temperature. With amorphous polymers orientation is produced at temperatures above the glass transition temperature. Below that temperature an amorphous polymer is too brittle and breaks. The effect of the chains becoming orientated is to give a harder, stronger material. The effect can be considered to be similar to work hardening with metals.

In order to improve the strength of polymer fibres, e.g. polyester fibres, they are put through a drawing operation to orientate the polymer chains. Stretching a polymer film causes orientation of the polymer chains in the direction of the stretching forces. The result is an increase in strength and stiffness in the stretching direction. Such stretching is referred to as uniaxial stretching and the effect as *uniaxial orientation*. The material is however weak and has a tendency to split if forces are applied in directions other than those of the stretching forces. For the polyester fibres this does not matter as the forces will be applied along the length of the fibre, but in film this could be a serious defect. The problem can be overcome by using a biaxial stretching process in which the film is stretched in two directions at right angles to each other. The film has then *biaxial orientation*.

Rolling through compression rollers is similar in effect to the drawing operation and results in an uniaxial orientated product. Extrusion has a similar effect.

Orientation can be obtained by both cold- and hot-working processes. Hot working involves working at temperatures just below the glass transition temperature. Under such conditions orientation can be produced without any internal stresses being developed. Cold working, at lower temperatures than those of hot working, requires more energy to produce orientation but also results in internal stresses being produced.

If orientated polymers are heated to above their glass transition temperature they lose their orientation. On cooling they are no longer orientated and are in the same state as they were before the orientation process occurred. This effect is made use of with shrinkable films. The polymer film is stretched, and so made longer even when the stretching force is removed. If it is then wrapped around some package and heated, the film contracts back to its initial, pre-stretched state. The result is a plastic film tightly fitting the package. The film is said to show *elastic memory*.

ADDITIVES

The term *plastics* is commonly used to describe materials based on polymers, other than elastomers, to which other substances have been added to give the required properties. Additives used fall into a number of categories:

1 Stabilisers to increase the degradation resistance of the material when in use.
2 Flame retardants to modify the fire properties of the material.
3 Fillers to increase such properties as tensile strength, hardness and impact strength and reduce thermal expansivity and cost.
4 Plasticisers to make the material softer and less rigid.
5 Antistats to reduce the electrostatic charging that can occur when plastic surfaces rub against each other.
6 Colorants to change the colour of the material.
7 Lubricants and heat stabilisers to assist processing.

The following are some examples of the additives that are commonly used with polymers.

Some plastics are damaged by ultraviolet radiation. Thus the effect of protracted periods of sunlight can lead to a deterioration of mechanical properties as well a reduction in transparency or change in colour. An ultraviolet absorber is thus often added to plastics, carbon black being often used. Such an additive is called a *stabiliser*.

Polymers contain the same basic ingredients, carbon and hydrogen, as fuels such as oil and gas. Thus given suitable conditions they can catch fire. Flame retardants, such as hydrogen bromide, can be incorporated within a plastic to inhibit the combustion reaction between the polymer and the oxygen in the air by itself combining with parts of the polymer.

The properties and the cost of a plastic can be markedly affected by the addition of other substances, these being termed *fillers*. The following table shows some of the common fillers used and their effects on the properties of the plastic.

Filler	*Effect on properties*
Asbestos	Improves temperature resistance, i.e. the plastic does not deform until a higher temperature is attained. Decreases strength and rigidity
Cotton flock	Increases impact strength but reduces electrical properties and water resistance
Cellulose fibres	Increases tensile strength and impact strength
Glass fibres	Increases tensile strength but lowers ductility. Makes the plastic stiffer
Mica	Improves electrical resistance
Graphite	Reduces friction
Wood flour	Increases tensile strength but reduces weater resistance

Where the filler improves the tensile strength it generally does so by reducing the mobility of the polymer chains. An important consideration however in the use of any filler is the fact that the fillers are generally cheaper than the polymer and thus reduce the overall cost of the plastic. Often up to 80% of a plastic may be filler.

One form of additive used is a gas. The result is foamed or 'expanded' plastics. Expanded polystyrene is used as a lightweight packing material. Polyurethanes in the expanded form are used as the filling for upholstery and as sponges.

An important group of additives are called *plasticisers*. Their primary purpose is to enable the molecular polymer chains to slide more easily past each other.

Internal plasticisation involves modifying the polymer chain by the introduction into the chain of bulky side groups. An example of this is the plasticisation of polyvinyl chloride by the inclusion of some 15% of vinyl acetate in the polymer chains. These bulky side groups have the effect of forcing the polymer chains further apart, so reducing the attractive forces between the chains and so permitting easier flow of chains past each other.

A more common method of plasticisation used is called *external plasticisation*. It involves a plasticiser being added to the polymer, after the chains have been produced. This plasticiser may be a liquid which disperses throughout the plastic, filling up the spaces between the chains. The effect of the liquid is the same as adding a lubricant between two metal surfaces, the polymer chains slide more easily past each other. The effect of the plasticiser is to weaken the attractive forces that exist between the polymer chains. The plasticiser decreases the crystallinity of polymers as it tends to hinder the formation of orderly arrays of polymer chains. The plasticiser also reduces the glass transition temperature. The effect on the mechanical properties is to reduce the tensile strength and increase the flexibility. The following table shows the effects of plasticiser on the mechanical properties of polyvinyl chloride.

	Tensile strength $/N\ mm^{-2}$ (MPa)	*Elongation* %
No plasticiser	52 to 58	2 to 40
Low plasticiser	28 to 42	200 to 250
High plasticiser	14 to 21	350 to 450

The PVC with no plasticiser is a rigid material, with low plasticiser content the PVC is pliable, with very high plasticiser content the PVC is soft and rubbery.

Plastics when rubbing against each other, or other materials, can readily become electrically charged. This shows itself as plastic films clinging together and being difficult to separate, and as dust 'sticking' to plastics. You can gain some idea of these effects by rubbing a piece of plastic, perhaps a ball-point casing, against a piece of cloth and then bringing it close to a small piece of paper which will be attracted and will 'stick to' the plastic. Such effects in the industrial processing of plastics can be very troublesome. For example, in handling photographic film, the last thing anybody wants is dust clinging to the film and ending up as small marks on the processed negatives. One way of reducing such effects is to add antistats to the polymer. They work by increasing the electrical conductivity of the plastic so that charges leak away rather than building up.

100 Polymeric materials

Pigments may be added to polymers to give a coloured plastic. Such materials act like fillers. Some pigments unfortunately have limited colour stability when exposed to sunlight and so the colour fades.

When PVC is softened during its processing it becomes sticky and so does not flow easily through the processing equipment. To ease its flow a lubricant can be added. Also to improve the thermal stability of a polymer during processing a heat stabiliser may be added. Such a stabiliser can also improve the thermal stability of the palstic during its service life.

COPOLYMERS

The term *homopolymer* is used to describe those polymers which are made up of just one monomer, e.g. polyethylene, made up of just the monomer ethylene. Other types of polymers can be produced by combining two or more monomers in a single polymer chain. *Figure 6.11* shows how two monomers can be combined to give different copolymer structures. Copolymerisation is used to improve the properties of polymers.

Polystyrene, with no additives, is a relatively hard, stiff and brittle material. Incorporating rubber particles, as an additive, in the polystyrene gives a material known as toughened polystyrene with marked improvements in impact resistance, though at the expense of a decrease in stiffness, i.e. tensile modulus, and tensile strength. Another way of improving the properties is to form a copolymer between polystyrene and acrylonitrile. This copolymer with an elastomer additive gives a very useful material known as ABS, acrylonitrile–butadiene–styrene. Such a material is tough, stiff and resists abrasion.

Elastomers are amorphous polymers with cross-links between the molecular chains (see later in this chapter). In the mid-1960s thermoplastic elastomers were developed for commercial use. An important example of these materials is the styrene–butadiene copolymer. The structure is a block copolymer, with alternating chains of polybutadiene and polystyrene. The polybutadiene blocks give the rubbery properties while the hard, glassy, polystyrene

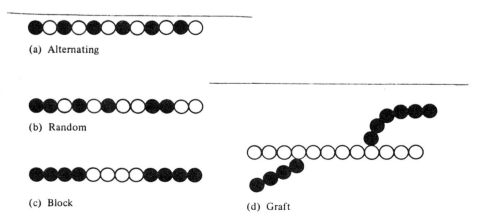

Figure 6.11 Structures of copolymers made up of two monomers

blocks give the cross-links between the chains. In addition they act as a reinforcing filler and increase the tensile modulus and strength. At temperatures above the glass transition temperature of polystyrene, 100°C, the material behaves like a thermoplastic and can be processed like a thermoplastic.

Low density polyethylene is partially crystalline and because of this is flexible and tough. High density polyethylene is more crystalline and as a result is stronger and stiffer but less tough. The copolymerisation of ethylene with vinyl acetate enables a material to be produced which is tougher than low density polythene and flexible, as a result of reducing the crystallinity, and gives a material having properties more like those of a rubber than a thermoplastic. The material ethylene-vinyl acetate, EVA, is used as a hot melt adhesive, for pipes, for tyres, for toy cars, and for applications in general similar to those of a rubber.

COMMON THERMOPLASTICS AND THEIR PROPERTIES

The following are brief outlines of the main properties and uses of thermoplastics commonly used in engineering (*Figure 6.12* and *6.13*):

Polyethylene (polythene) is referred to as high density when the linear chain polymer is involved and low density when the branched form is used. The high density polyethylene has the greater crystallinity.

Property	Low density polyethylene	High density polyethylene
Crystallinity	60%	95%
Density/10^3 kg m^{-3}	0.92	0.95
T_g/°C	−120	−120
T_m/°C	115	138
Tensile strength/N mm^{-2} (MPa)	8 to 16	22 to 38
Tensile modulus/kN mm^{-2} (GPa)	0.1 to 0.3	0.4 to 1.3
Elongation %	100 to 600	50 to 800

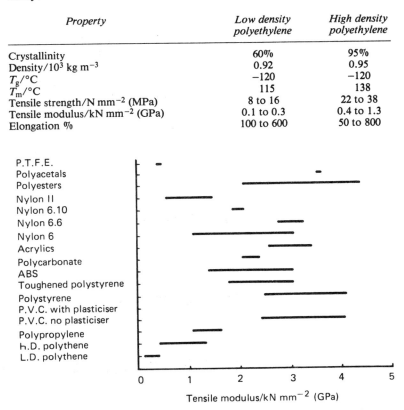

Figure 6.12 Comparison of tensile modulus values for thermoplastics

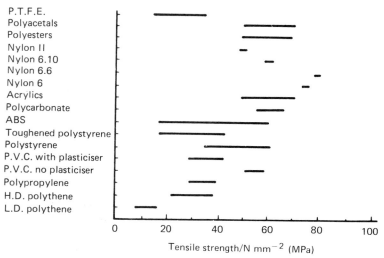

Figure 6.13 Comparison of tensile strength values for thermoplastics

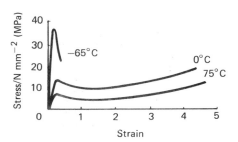

Figure 6.14 Stress–strain graphs for low density polythene

Figure 6.15 Stress–strain graph for polypropylene

Low- and high-density polymers may be blended to give plastics with properties intermediate between those quoted above. *Figure 6.14* shows the stress–strain graph for low-density polythene.

Low-density polyethylene is used mainly in the form of films and sheeting, e.g. polythene bags, 'squeeze' bottles, ball-point pen tubing, wire and cable insulation. High-density polythene is used for piping, toys, filaments for fabrics, household ware. Both forms have excellent chemical resistance, low moisture absorption and high electrical resistance. The additives commonly used with polyethylene are carbon black as a stabiliser, pigments to give coloured forms, glass fibres to give increased strength and butyl rubber to prevent inservice cracking.

Polypropylene is used mainly in its crystalline form, being a linear polymer with side groups regularly arranged along the chain. The presence of these side groups give a more rigid and stronger polymer (*Figure 6.15*) than polyethylene in its linear form.

Property	Polypropylene
Crystallinity	60%
Density/10^3 kg m^{-3}	0.90
T_g/°C	-10
T_m/°C	176
Tensile strength/N mm^{-2} (MPa)	30 to 40
Tensile modulus/kN mm^{-2} (GPa)	1.1 to 1.6
Elongation %	50 to 600

Polypropylene is used for crates, containers, fans, car fascia panels, tops of washing machines, cabinets for radios and televisions, toys, chair shells.

Polyvinyl chloride (PVC) is a linear chain polymer with bulky side groups (see *Figure 6.4*) which prevent crystalline regions occurring. The material is hard and rigid. It is often used with a plasticiser to give a more flexible plastic.

Property	With no plasticiser	Low plasticiser content
Crystallinity	0	0
Density/10^3 kg m^{-3}	1.4	1.3
T_g/°C	87	—
T_m/°C	—	—
Tensile strength/N mm^{-2}	52 to 58	28 to 42
Tensile modulus/kN mm^{-2}	2.4 to 4.1	—
Elongation %	2 to 40	200 to 250

The rigid form of PVC, i.e. the unplasticised form, is used for piping for waste and soil drainage systems, rainwater pipes, lighting fittings, curtain rails. Plasticised PVC is used for the fabric of 'plastic' raincoats, bottles, shoe soles, garden hose piping, gaskets, inflatable toys.

Polystyrene is a linear chain polymer with bulky side groups which prevent crystalline regions occuring. Polystyrene with no additives is a brittle, transparent material. A toughened form of polystyrene is produced by blending polystyrene with rubber particles.

Property	No additives	Toughened
Crystallinity	0	0
Density/10^3 kg m^{-3}	1.1	1.1
T_g/°C	100	—
T_m/°C	—	—
Tensile strength/N mm^{-2} (MPa)	35 to 60	17 to 42
Tensile modulus/kN mm^{-2} (GPa)	2.5 to 4.1	1.8 to 3.1
Elongation %	1 to 3	8 to 50

Polystyrene is an excellent electrical insulator and is widely used in electrical equipment. Other applications are packaging of cosmetics, toys, boxes. A widely used form of polystyrene is as expanded polystyrene, which is a rigid foam used widely for insulation and packaging. Toughened polystyrene is less brittle than ordinary polystyrene and finds use as cups in vending machines, casing for cameras, projectors, radios, television sets, vacuum cleaners.

Acrylonitrile–butadiene–styrene terpolymer (ABS) is produced by forming polymer chains with three different polymer materials, polystyrene, acrylonitrile and butadiene. It gives an amorphous material which is tough, stiff and abrasion resistant.

Property	ABS
Crystallinity	0
Density/10^3 kg m^{-3}	1.1
T_g/°C	71 to 112
T_m/°C	—
Tensile strength/N mm^{-2} (MPa)	17 to 58
Tensile modulus/kN mm^{-2} (GPa)	1.4 to 3.1
Elongation %	10 to 140

ABS is widely used as the casing for telephones, vacuum cleaners, hair driers, radios, television, typewriters, luggage, boat shells, food containers.

Polycarbonate is an amorphous thermoplastic having a linear chain with bulky side groups. It is tough, stiff and strong.

Property	Polycarbonate
Crystallinity	0
Density/10^3 kg m^{-3}	1.2
T_g/°C	150
T_m/°C	—
Tensile strength/N mm^{-2} (MPa)	55 to 65
Tensile modulus/kN mm^{-2} (GPa)	2.1 to 2.4
Elongation %	60 to 100

It is used for applications where the plastics required are resistant to impact abuse and, for plastics, relatively high temperatures. Typical applications are transparent street lamp covers, infant-feeding bottles, machine housings, safety helmets, housings for car lights, and tableware such as cups and saucers.

Acrylics are completely transparent thermoplastics having linear chains with bulky side groups and so giving an amorphous structure. They give a stiff, strong material with outstanding weather resistance.

Property	Acrylics
Crystallinity	0
Density/10^3 kg m^{-3}	1.18
T_g/°C	0
T_m/°C	—
Tensile strength/N mm^{-2} (MPa)	50 to 70
Tensile modulus/kN mm^{-2} (GPa)	2.7 to 3.5
Elongation %	5 to 8

Because of its transparency acrylic is used for windscreens, light fittings, canopies, lenses for car lights, signs and nameplates. Opaque acrylic sheet is used for the production of domestic baths, shower cabinets, basins and lavatory cisterns.

Polyamides, or nylons as they are better known, are linear polymers and give crystalline structures. There are a number of common polyamides: nylon 6, nylon 6.6, nylon 6.10 and nylon 11. The numbers refer to the numbers of carbon atoms in each of the reacting substances used to give the polymer. The full stops separating the two numbers are sometimes omitted, e.g. nylon 66 is nylon 6.6. The two most used nylons are nylon 6 and nylon 6.6. Nylon 6.6 has a higher melting point than nylon 6 and is also stronger and stiffer. Nylon 11 has a lower melting point than either nylon 6 or 6.6; it is also more flexible. Nylon 6.10 has properties intermediate between nylon 11 and nylon 6.6.

In general, nylon materials are strong, tough and have relatively high melting points. But they do tend to absorb moisture, the effect of which is to reduce their tensile strength. Nylon 6.6 can absorb quite large amounts of moisture, nylon 11 however absorbs considerably less.

Property	Nylon 6	Nylon 6.6	Nylon 6.10	Nylon 11
Crystallinity	Can be varied from low to high percentages			
Density/10^3 kg m^{-3}	1.13	1.1	1.1	1.1
T_g/°C	50	57	50	—
T_m/°C	225	265	228	—
Tensile strength/N mm^{-2} (MPa)	75	80	60	50
Tensile modulus/kN mm^{-2} (GPa)	1.1 to 3.1	2.8 to 3.3	1.9 to 2.1	0.6 to 1.5
Elongation %	60 to 320	60 to 300	85 to 230	70 to 300

Nylons often contain additives, e.g. a stabiliser or flame retardant substance. Glass spheres or glass fibres are added to give improved strength and rigidity. Molybdenum disulphide is an additive to nylon 6 to give a material with very low frictional properties.

Nylon is used for the manufacture of fibres for clothing, gears, bearings, brushes, housings for domestic and power tools, electric plugs and sockets.

Polyesters are available in a thermoplastic form, usually *polyethylene terephthalate*. This is a linear chain polymer with side groups. It gives crystalline structures and is below its glass transition point at normal temperatures. If it is rapidly quenched from a melt to below its glass transition point an amorphous structure is produced, the molecular chains not having sufficient time to become packed in an orderly way.

Property	Polyethylene terephthalate
Crystallinity	60%
Density/10^3 kg m^{-3}	1.3
T_g/°C	69
T_m/°C	267
Tensile strength/N mm^{-2} (MPa)	50 to 70
Tensile modulus/kN mm^{-2} (GPa)	2.1 to 4.4
Elongation %	60 to 100

The polyester has properties similar to nylon. It is used widely in fibre form for the production of clothes. Other uses are for electrical plugs and sockets, push-button switches, wire insulation, recording tapes, insulating tapes, gaskets.

Polyacetals, polyformaldehyde or polyoxymethylene, are linear polymers giving rise to crystalline structures. They are stiff, strong polymers and maintain their properties at relatively high temperatures. They are however adversely affected by ultraviolet light and thus have to be used with a stabiliser.

Property	Polyformaldehyde	Polyoxymethylene
Crystallinity	High	High
Density/10^3 kg m^{-3}	1.41	1.41
T_g/°C	−73	−76
T_m/°C	180	180
Tensile strength/N mm^{-2} (MPa)	50 to 70	70
Tensile modulus/kN mm^{-2} (GPa)	3.6	3.6
Elongation %	15 to 75	15 to 75

Typical applications are pipe fittings, parts for water pumps and washing machines, car instrument housings, bearings, gears, hinges and window catches, seat belt buckles.

Polytetrafluoroethylene (PTFE) is a linear polymer like polyethylene, the only difference being that instead of hydrogen atoms there are fluorine atoms. It has a very high crystallinity as manufactured, about 90%, though this degree of crystallinity can be reduced during processing to about 50% if quench cooled, or 75% with slow cooling. Tough and flexible, PTFE can be used over a wide range of temperature, 250°C down to almost absolute zero, and still retain the very important property of not being attacked by any reagent or solvent. It also has a very low coefficient of friction. No known material can be used to bond it satisfactorily to other materials.

Property	PTFE
Crystallinity	90%
Density/10^3 kg m^{-3}	2.2
T_g/°C	−120
T_m/°C	327
Tensile strength/N mm^{-2} (MPa)	14 to 35
Tensile modulus/kN mm^{-2} (GPa)	0.4
Elongation %	200 to 600

PTFE is a relatively expensive material and is not processed as easily as other thermoplastics. It tends to be used where its special properties, i.e. resistance to chemical attack and very low coefficient of friction, are needed. Journal bearings with a PTFE surface can be used without lubrication because of the low coefficient of friction; they can even be used at temperatures up to about 250°C. Piping carrying corrosive chemicals at temperatures up to 250°C are made of PTFE. Other applications for PTFE are gaskets, diaphragms, valves, O-rings, bellows, couplings, dry and self-lubricating bearings, coatings for frying pans and other cooking utensils (known as 'non-stick'), coverings for rollers handling sticky materials, linings for hoppers and chutes, and electrical insulating tape.

THERMOSETTING POLYMERS

Thermoplastic polymers soften if heated, and can be reshaped, the new shape being retained when the plastic cools. The process can be repeated. *Thermosetting polymers* cannot be softened and reshaped by heating. They are plastic in the initial stage of manufacture but once they have set they cannot be resoftened. The atoms in a thermosetting material form a three-dimensional structure of cross-linked chains (see *Figure 6.1c*). The bonds linking the chains are strong and not easily broken. Thus the chains cannot slide over one another but are essentially fixed in the positions they occupied when the polymer was solidifying during its formation.

Thermosetting polymers are stronger and stiffer than thermoplastics and generally they can be used at higher temperatures than thermoplastics. As they cannot be shaped after the initial reaction in which the polymer chains are produced, the processes by which thermosetting polymers can be shaped are limited to those where the product is formed from the raw polymer materials. No further processing is possible (other than possibly some machining) and this limits the processes available to essentially just *moulding*. A number of different moulding methods are used but essentially all involve the combining together of the chemicals in a mould so that the cross-linked chains are produced while the material is in the mould. The result is a thermosetting polymer shaped to the form dictated by the mould.

Properties of common thermosets

Phenolics give highly cross-linked polymers. *Phenol formaldehyde* was the first synthetic plastic and is known as *Bakelite*. The polymer is opaque and initially light in colour. It does however darken with time and so is always mixed with dark pigments to give dark-coloured materials. It is supplied in the form of a moulding powder, including the resin, fillers and other additives such as pigments. When this moulding powder is heated in a mould the cross-linked polymer chain structure is produced. The fillers account for some 50 to 80% of the total weight of the moulding powder. Wood flour, a very fine soft wood sawdust, when used as a filler increases the impact strength of the plastic, asbestos fibres improve the heat properties, and mica the electrical resistance.

Property	Unfilled	Wood flour filler	Asbestos filler
Density/10^3 kg m^{-3}	1.25 to 1.30	1.32 to 1.45	1.6 to 1.85
Tensile strength/N mm^{-2} (MPa)	35 to 55	40 to 55	30 to 55
Tensile modulus/kN mm^{-2} (GPa)	5.2 to 7.0	5.5 to 8.0	0.1 to 11.5
Elongation %	1 to 1.5	0.5 to 1	0.1 to 0.2
Max. service temp./°C	120	150	180

Phenol formaldehyde mouldings are used for electrical plugs and sockets, switches, door knobs and handles, camera bodies and ash trays. Composite materials involving the phenolic resin being used with paper or an open weave fabric, e.g. a glass fibre fabric, are used for gears, bearings, and electrical insulation parts.

Amino-formaldehyde materials, generally *urea formaldehyde* and *melamine formaldehyde*, give highly cross-linked polymers. Both are used as moulding powders, like the phenolics. Cellulose

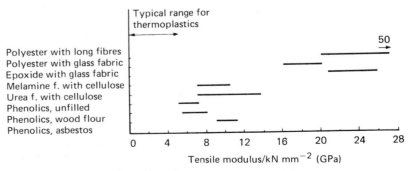

Figure 6.16 Comparison of tensile modulus values of thermosets

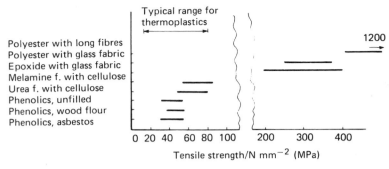

Figure 6.17 Comparison of tensile strength values of thermosets

and wood flour are widely used as fillers. Hard, rigid, high-strength materials are produced.

Property	Urea formaldehyde cellulose filler	Melamine formaldehyde cellulose filler
Density/10^3 kg m^{-3}	1.5 to 1.6	1.5 to 1.6
Tensile strength/N mm^{-2} (MPa)	50 to 80	55 to 85
Tensile modulus/kN mm^{-2} (GPa)	7.0 to 13.5	7.0 to 10.5
Elongation %	0.5 to 1.0	0.5 to 1.0
Max. service temp./°C	80	95

Both materials are used for table ware, e.g. cups and saucers, knobs, handles, light fittings and toys. Composites with open weave fabrics are used as building panels and electrical equipment.

Epoxide materials are generally used in conjunction with glass, or other, fibres to give hard and strong composites. Epoxy resins are excellent adhesives giving very high adhesive strengths.

Property	Plain weave glass fabric
Percentage fabric	60 to 65%
Density/10^3 kg m^{-3}	1.8
Tensile strength/N mm^{-2} (MPa)	200 to 420
Tensile modulus/kN mm^{-2} (GPa)	21 to 25
Max. service temp./°C	200

The unfilled epoxide has a tensile strength of 35 to 80 N mm^{-2} (MPa), considerably less than that of the composite. The composite is used for boat hulls and table tops.

Polyesters can be produced as either thermosets or thermoplastics. The thermoset form is mainly used with glass or other fibres to form hard and strong composites. Such composites are used for boat hulls, architectural panels, car bodies, panels in aircraft, and stackable chairs. They have a maximum service temperature of the order of 200°C.

ELASTOMERS

Elastomers are polymers which can show very large, reversible, strains when subjected to stress. *Figure 6.18* shows a typical stress–strain graph for an elastomer. As the graph indicates, Hooke's law

Polymeric materials 109

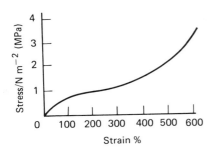

Figure 6.18 Typical stress–strain graph for an elastomer

is not obeyed and the tensile modulus increases at large strains. The behaviour of the material is perfectly elastic up to considerable strains, e.g. you can stretch a rubber band up to more than five times its unstrained length and it still is elastic.

Elastomers are essentially amorphous polymers with a glass transition temperature below their service temperature. The polymer structure is that of linear chain molecules with some cross-linking between chains. A simple model for the structure might be a piece of netting. In the unstretched state the netting is in a loose pile, thus when the material begins to be stretched the netting just begins to untangle itself and large strains are easily produced. This explains the low value of the tensile modulus that occurs with elastomers and the large strains that are possible. It is not until quite large strains, when the netting has assumed an orderly arrangement, that the bonds between atoms in the material begin to be significantly stretched; then the stress–strain graph shows an increase in tensile modulus. *Figure 6.19* shows this simple model of an elastomer in the initially unstretched state and when at a large strain.

Properties of common elastomers

Natural rubber is, in its crude form, just the sap from a particular tree. The addition of sulphur to the rubber produces cross-links, the amount of cross-linkage being determined by the amount of sulphur added. The process of producing cross-links is called *vulcanisation*. Anti-oxidants and plasticisers are also added to the rubber.

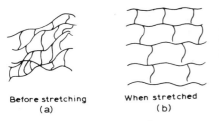

Figure 6.19 Stretching an elastomer which has crosslinks

Property	Natural rubber
Tensile strength/N mm^{-2} (MPa)	20
Elongation %	800
Service temp. range/°C	−50 to +80
Resistance to oils and greases	Poor
Resistance to water	Good

Butadiene styrene rubbers, commonly called *SBR* or *GR-S* or *Buna S* rubbers, are synthetic rubbers. They are cheaper than natural rubber. They are used in the manufacture of tyres, hosepipes, conveyor belts and cable insulation.

Property	SBR
Tensile strength/N mm^{-2} (MPa)	24
Elongation %	600
Service temp. range/°C	−50 to +80
Resistance to oils and greases	Poor
Resistance to water	Good

Butyl rubbers, known as *isobutylene isoprene* or *GR-I*, have the important property of extreme impermeability to gases. They are thus widely used for the inner linings of tubeless tyres, steam hoses and diaphragms.

Property	Butyl
Tensile strength/N mm^{-2} (MPa)	20
Elongation %	900
Service temp. range/°C	−50 to +100
Resistance to oils and greases	Poor
Resistance to water	Good

Nitrile rubbers, known as *butadiene acrylonitrile* or *Buna N*, are extremely resistant to organic liquids and used for such applications as hoses, gaskets, seals, tank linings, rollers, valves.

Property	Butyl
Tensile strength/N mm^{-2} (MPa)	28
Elongation %	700
Service temp. range/°C	−50 to +125
Resistance to oils and greases	Excellent
Resistance to water	Good

Neoprene, known as *polychloroprene*, has good resistance to oils and a variety of other chemicals, also having good weathering characteristics. It is used for oil and petrol hoses, gaskets, seals, diaphragms and chemical tank linings.

Property	Neoprene
Tensile strength/N mm^{-2} (MPa)	25
Elongation %	1000
Service temp. range/°C	−50 to +100
Resistance to oils and greases	Good
Resistance to water	Fair

Polyurethane rubbers have higher tensile strengths, tear and abrasion resistance than other rubbers. They are relatively hard and offer good resistance to oxygen and ozone. Their main uses are in oil seals, diaphragms, tyres of forklift trucks or other vehicles where low speeds are involved (not high speeds since they have a low skid resistance), heels and soles of shoes and industrial chute linings.

Property	Polyurethane
Tensile strength/N mm^{-2} (MPa)	36
Elongation %	650
Service temp. range/°C	−55 to +125
Resistance to oils and greases	Good
Resistance to water	Good

Silicone rubbers are based on chains of silicon atoms rather than carbon atoms, as with the other elastomers. Although silicone rubbers have, compared with other rubbers, relatively low tensile strength, they retain their rubber-like properties and are resistant to oils, fuels and water over a very wide temperature range. They are however considerably more expensive than carbon based elastomers. Because of this their use tends to be restricted to parts requiring their properties at either high or low temperatures, e.g. seals, gaskets, O-rings and insulation for wires and cables.

Property	Silicone rubber
Tensile strength/N mm^{-2} (MPa)	10
Elongation %	700
Service temp. range/°C	−105 to 300
Resistance to oils and greases	Fair
Resistance to water	Excellent

POLYMER FOAMS

Foamed versions of many plastics and elastomers have been developed, e.g. polystyrene, polyethylene, urea formaldehyde, styrene butadiene rubber. The processes used for producing the foams are:

1 Mechanical agitation to disperse air through the liquid polymer. This generally produces materials with an open-cell structure.

2 Gas bubbles are formed within the liquid polymer, by heating or a reduction in pressure or as a result of chemical reaction. This can produce materials with an open or closed cell structure, depending on the conditions prevailing.

The flexibility or stiffness of a foam depends on whether it has open or closed cells. Closed cells give a more rigid structure as a consequence of the air trapped in the cells.

Flexible foams are widely used as cushioning material. Rigid foams are used in sandwich panel composites with metal or sheet polymer facing sheets, for thermal insulation and packaging.

POLYMER BASED SANDWICH MATERIALS

Sandwich materials are a form of laminated composite. Such materials can be made of metal or polymer sheets sandwiching a core of a metal honeycomb or a foamed plastic. The core forms the bulk of the composite material. The result of such a combination is a material with a high stiffness, as a result of the facing sheets, and low overall density. Such materials are frequently used where rigidity with low density is required, e.g. internal panelling in aircraft.

POLYMER COMPOSITES

Reinforced plastics consist of a stiff, strong material combined with the plastic. Glass fibres are probably the most used additive. The fibres may be long lengths, running through the length of the composite, or discontinuous short lengths randomly orientated within the composite. Another form of composite uses glass fibre mats or cloth in the plastic. The effect of the additives is to increase both the tensile strength and the tensile modulus of the plastic, the amount of change depending on both the form the additive takes and the amount of it. The continuous fibres give the highest tensile modulus and tensile strength composite but with a high directionality of properties. The strength along the direction of the fibres could be perhaps 800 N mm^{-2} (MPa) while that at right-angles to the fibre direction may be as low as 30 N mm^{-2} (MPa), i.e. just about the strength of the plastic alone. Randomly orientated short fibres do not lead to this directionality of properties but do not give such high strength and tensile modulus. The composites with glass fibre mats or cloth tend to give tensile strength and modulus values intermediate between those of the continuous and short length fibres. The following are examples of the strength and modulus values obtained with reinforced polyester.

Material	Percentage weight of glass	Tensile modulus /kN mm^{-2} (GPa)	Tensile strength /N mm^{-2} (MPa)
Polyester	0	2 to 4	20 to 70
With short fibres	10 to 45	5 to 14	40 to 180
With plain weave cloth	45 to 65	10 to 20	250 to 350
With long fibres	50 to 80	20 to 50	400 to 1200

PROBLEMS

1 How is the structure of an amorphous polymer different from that of a crystalline polymer?

2 How does the form of the polymer molecular chain determine the degree of crystallinity possible with the polymer?

3 Explain how the crystallinity of a polymer affects its properties.

4 What is the glass transition temperature?

5 PVC has a glass transition temperature of 87°C. How would its properties below this temperature differ from those above it?

6 A polypropylene article was designed for use at room temperature. What difference might be expected in its behaviour if used at about −15°C? The glass transition temperature for polypropylene is −10°C.

7 When a piece of polythene is pulled it starts necking at one point. Further pulling results in no further reduction in the cross-section of the material at the necked region but a spread of the necked region along the entire length of the material. Why doesn't the material just break at the initial necked section instead of the necking continuing?

8 Why are the properties of a cold drawn polymer different from those of the undrawn polymer?

9 Why are polyester fibres cold drawn before use?

10 Polypropylene is a crystalline polymer with a glass transition temperature of −10°C and a melting point of 176°C. What would be its normal maximum service temperature? At what temperatures would the polymer be hot formed?

11 Explain what is meant by internal and external plasticisation.

12 What is the effect of a plasticiser on the mechanical properties of a polymer?

13 Describe how the engineering properties of polymers may be changed by incorporating additives into the materials.

14 Explain what is meant by copolymerisation.

15 Describe how the properties of polystyrene are modified by copolymerisation with acrylonitrile.

16 How do the properties of high and low density polythene differ?

17 How do the properties of polystyrene with no additives and toughened polystyrene differ?

18 To what temperature would high density polythene have to be heated to be hot formed?

19 Which is stiffer at room temperature, PVC with no plasticiser or polypropylene?

20 The casing of a telephone is made from ABS. How would the casing behave if someone left a burning match or cigarette against it?

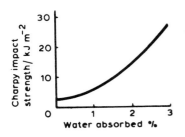

Figure 6.20 The effect of water absorption on the impact strength of nylon 6

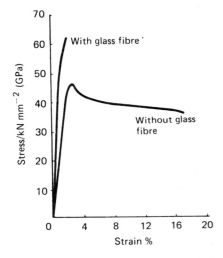

Figure 6.21 Stress–strain graphs for ABS plastics Novodur PHGV (reinforced) and PH-AT (unreinforced) (courtesy of Bayer UK Ltd)

21 What are the special properties of PTFE which render it useful despite its high price and processing problems?

22 How do the mechanical properties of thermosets differ, in general, from those of thermoplastics?

23 What is bakelite and what are its mechanical properties?

24 Describe how cups made of melamine formaldehyde with a cellulose filler might be expected to behave in service.

25 What is meant by vulcanisation?

26 Would Buna S or Buna N be the best rubber to use for a pipe carrying oil?

27 Explain how rubbers can be stretched to several times their length and return to the same initial length when released.

28 *Figure 6.20* shows the effect on the Charpy impact strength for nylon 6 of percentage of water absorbed. As the percentage of water absorbed increases, is the material becoming more or less brittle?

29 *Figure 6.21* shows the stress–strain graphs for two forms of ABS plastic, one containing 20% by weight glass fibre and the other without such fibres. Estimate from the graph (a) the tensile strength and (b) the tensile modulus for both forms.

30 Which polymers would be suitable for the following applications? (a) An ash tray. (b) A garden hose pipe. (c) A steam hose. (d) Insulation for electric wires. (e) The transparent top of an electrical meter. (f) A plastic raincoat. (g) A toothbrush. (h) A camera body.

31 Describe typical applications of (a) polymer foams, (b) polymer based sandwich materials.

Index

ABS, 100, 102
Acrylics, 104
Acrylonitrile-butadiene-styrene terpolymer, 100, 102
Additives, polymer, 98
Ageing, 87
Alloy steels, 40, 56
Alloys, 39
Aluminium, 71
Aluminium alloys, 72
Amino-formaldehyde, 107
Amorphous structures, 92
Annealing, 59
Antistats, 99
Arrest points, 58
Austenite, 53
Austenitic steel, 57

Bakelite, 107
Brasses, 41, 78
Brinell test, 9
Brittleness, 1, 9, 16, 31
Bronzes, 41, 79
Butadiene-styrene rubber, 109
Butyl rubber, 109

Carbon steel, 40
Carbonitriding, 67
Carburising, 65
Case depth, 67
Case hardening, 65
Cementite, 53
Charpy test, 15
Composites, 111
Compound, 42
Copolymer, 100
Copper, 41, 77
Copper alloys, 41, 77
Creep, 22, 33
Creep modulus, 33
Critical change points, 58
Critical cooling rate, 57, 60
Crystallinity, polymers, 92
Cyaniding, 66

Ductility, 1, 9, 16, 31
Duralumin, 73

Elastic limit, 2
Elasticity, 1
Elastomers, 91, 108
Endurance limit, 20
Epoxide, 108

Equilibrium diagrams, 44
Ethylene-vinyl acetate, 101
Eutectic, 46
Eutectoid, 53

Fatigue limit, 19
Fatigue strength, 1
Fatigue tests, 19, 35
Ferrite, 53
Ferritic nitrocarburising, 67
Ferritic steel, 57
Ferrous alloys, 52
Fillers, 98
Flame hardening, 65
Foams, polymer, 111
Furnace:
    muffle, 68
    salt bath, 69

Glass transition temperature, 94

Hardenability, 64
Hardening, surface, 65
Hardness, 1, 9
    measurement 9, 34
Heat stabiliser, 100
Heat treatment, 59
High carbon steel, 50
Homopolymer, 100
Hooke's law, 2
Hyper-eutectoid steel, 54
Hypo-eutectoid steel, 54

Impact tests, 14, 34
Induction hardening, 65
Iron alloys, 39
Iron-carbon system, 52
Isochronous stress-strain graph, 33
Izod test, 15

Latent heat, 44
Limit of proportionality, 2
Limiting ruling section, 7
Liquidus, 46

Magnesium, 82
Magnesium alloys, 82
Malleability, 1
Martensite, 56, 60
Mass effect, 61, 65
Medium carbon steel, 50
Melamine formaldehyde, 107
Mild steel, 50

Mixture, 42
Moh scale, 13
Muffle furnace, 68

Neck, 2
Neoprene, 110
Nickel, 83
Nickel alloys, 82
Nickel silver, 41, 80
Nitriding, 66
Nitrile rubber, 110
Non-ferrous alloys, 71
Normalising, 60
Nylon, 104

Orientation, polymers, 32, 96

Pearlite, 54
Percentage elongation, 3
Percentage reduction in area, 4
Permanent set, 2
Phase, 43
Phenolics, 107
Plastic deformation, 2
Plasticisation, 99
Plasticity, 1
Plastics, 29
Polyacetals, 105
Polyamides, 104
Polycarbonate, 104
Polyesters, 105, 108
Polyethylene, 94, 101
Polyethylene terephthalate, 105
Polymers, 29, 91
Polypropylene, 102
Polystyrene, 102
Polytetrafluorethylene, 106
Polyurethane, 110
Polyvinylchloride, 92, 99, 102
Precipitation hardening, 73, 87
Process annealing, 60
Proof stress, 3
Proportional test piece, 5
PTFE, 106
PVC, 92, 99, 102

Rockwell test, 11, 34
Rubber, 109

Ruling section, 7, 65
Rupture stress, 23, 33

Salt bath furnace, 69
Sandwich materials, 111
Secant modulus, 30
Shore durometer, 34
Silicone rubber, 110
S/N graph, 19
Softness number, 34
Solid solutions, 42
Solidus, 46
Solubility, 42
Solution, 42
Solvus, 48
Spherodising annealing, 60
Stabilisers, 98
Stainless steel, 57
Strain, 2
Strength, 1, 18
Stress, 2
  rupture, 23, 33
Stress relief, 62
Subcritical annealing, 60
Surface hardening, 65

Temper, 71
Tempering, 61
Tensile modulus, 3
Tensile strength, 3
Tensile test, 1
Tensile test piece, 5
Thermal equilibrium diagram, 44
Thermoplastics, 91
Thermosets, 91, 106
Titanium, 83
Titanium alloys, 84
Toughness, 1, 18

Urea formaldehyde, 107

Vickers test, 10

Yield point, 3
Young's modulus, 3

Zinc, 84
Zinc alloys, 84